D1247871

André and Eveline Weil (photo by Lucien Gillet, May 2, 1948)

André Weil
The Apprenticeship
of a Mathematician

Translated from the French
by Jennifer Gage

1992   Birkhäuser Verlag
Basel · Boston · Berlin

Author's address

Prof. Dr. André Weil
Institute for Advanced Study
School of Mathematics
Princeton, NJ 08540, USA

This work was originally published as:
Souvenirs d'apprentissage; Basel: Birkhäuser 1991.

All illustrations were kindly provided by Professor A. Weil, Princeton.

Library of Congress Cataloging-in-Publication Data
Weil, André, 1906–
    [Souvenirs d'apprentissage. English]
    The apprenticeship of a mathematician / André Weil; translated
from the French by Jennifer Gage.
    Translation of: Souvenirs d'apprentissage.
    Includes index.
    ISBN 3-7643-2650-6 – ISBN 0-8176-2650-6
    1. Weil, André, 1906–   .   2. Mathematicians-France-Biography.
I. Title
QA29.W455A3   1992
510'.92 – dc20

Deutsche Bibliothek Cataloging-in-Publication Data
Weil, André:
The apprenticeship of mathematician / André Weil. Transl.
from the French by Jennifer Gage. - Basel ; Boston ; Berlin :
Birkhäuser, 1992
    Franz. Ausg. u.d.T.: Weil, André: Souvenirs d'apprentissage
    ISBN 3-7643-2650-6 (Basel ...)
    ISBN 0-8176-2650-6

© 1992 Birkhäuser Verlag Basel
Printed on acid-free paper in Germany
ISBN 3-7643-2650-6
ISBN 0-8176-2650-6

Translator's acknowledgments

I am grateful for the continuing support and encouragement of Rosanna Warren. For assistance with mathematical terminology my thanks go to Thanasis Kehagias – a true scion of Bourbaki. Finally, I am deeply indebted to Dr. Weil himself for his patient and generous assistance.

J.G.

Table of Contents

CONIUGIS DILECTAE
MANIBUS

Pero nadie querrá mirar tus ojos
porque te has muerto para siempre.

Federico García Lorca

Foreword

My life, or at least what deserves this name – a singularly happy life, its diverse vicissitudes withal – is bounded by my birth on May 6, 1906, and the death on May 24, 1986, of my wife and companion Eveline. If she figures rather little in these pages which I dedicate to her, it is not that she did not count for much in my life and thought; on the contrary, she was so intimately a part of these, almost from our very first meeting, that to speak of myself is also to speak of her: her presence and her absence form the warp on which the entire weft of my life was woven. What shall I say, but that our marriage was one of those which give the lie to La Rochefoucauld?[1] *Fulsere vere candidi mihi soles...*

My sister, too, is not much mentioned in these memoirs. Some time ago, in any case, I recounted my memories of her to Simone Pétrement, who recorded them in her fine biography, *Simone Weil: A Life*;[2] it would be superfluous to repeat herein those many details on our childhood which are included in Pétrement's book. As children, Simone and I were inseparable; but I was always the big brother and she the little sister. Later on we saw each other only rarely, speaking to one another most often in a humorous vein; she was naturally bright and full of mirth, as those who knew her have attested, and she retained her sense of humor even when the world had added on a layer of inexorable sadness. In truth we had few serious conversations. But if the joys and sorrows of her adolescence were never known to me at all, if her behavior later on often struck me (and probably for good cause) as flying in the face of common sense, still we remained always close enough to one another so that nothing about her really came as a surprise to me – with the sole exception of her death. This I did not expect, for I confess that I had thought her indestructible. It was not until quite late that I came to understand that her life had unfolded according to its own laws, and thus also did it end. I was little more than a distant observer of her trajectory.

All I am attempting here is to retrace the intellectual itinerary of a mathematician – a mathematician who has perhaps become too long-winded for the amusement (benevolent, I hope) of the younger generation. In the case of a writer or an artist, apparently nothing matters more than to

---

1   *Maximes* 113: "Il y a de bons mariages, mais il n'y en a point de délicieux" (there are good marriages, but there are no delicious ones).

2   Translated by Raymond Rosenthal (New York: Pantheon, 1976).

scrutinize the first gurglings of his infancy, after which the modern reader expects to be admitted into the most intimate recesses of the character's love life. But I have neither the temperament nor the talent of a Jean-Jacques Rousseau; and his is not the way to account for the life of a mathematician.

I had first planned to end these memoirs with my arrival in New York harbor in March of 1941 with my wife and her son Alain. But my years of apprenticeship (my *Lehrjahre*, which were also *Wanderjahre*) did not end so early. What am I saying? I am still learning today: I am learning to live in my memories. May the kindly reader accompany me with all good will... his presence will be precious to me.

Chapter I
Growing Up

I almost never recall my dreams, and my memory for faces could hardly be worse. Is it for the same reason that I have very few memories of childhood? Indeed, I have always imagined that the human memory has limited storage space, and that the art of memory is no less the art of forgetting than that of remembering.

My childhood memories, therefore, are fragmented and sparse, and they begin late. Besides, can we ever distinguish for sure between what is truly recollected and stories we have been told at a later date, which are unconsciously transformed into false remembrances? I seem to see in my mind a blurry picture of the floods that turned the Paris streets near the Seine into rivers in the winter of 1910. It is indeed possible that I was taken to view this unwonted spectacle; but perhaps this image is simply transposed from a description I later heard, or from a photograph or drawing in some magazine.

As for my family origins, I have only vague and patchy notions: the search for my "roots," as the current expression goes, never appealed to me. I knew neither of my grandfathers. One of them, Abraham Weill, died in Strasbourg before I was born. It seems that, as a respected member of the Jewish community in Alsace, he was often called upon to settle differences among others. In keeping with Jewish custom, after his first wife died he married her sister, and with her had two sons, my father and then my uncle Oscar. When I was born, my grandmother and my uncles had long since moved to Paris with their families, having exercised their right as Alsatians to opt for French nationality. My father, Bernard Weil (somewhere along the way, an *l* was lost) had done likewise, and was married in 1905.

Until 1912, my family lived at Number 19 on the Boulevard de Strasbourg, not far from the Gare de l'Est. One day my father, taking a walk with me along the boulevard, told me that my first name came from the Greek word meaning "man," and that this was one reason it had been given me. Did he go on to say that I must prove worthy of this name? I do not recall; but certainly that was the intention of his words, and it is thus that the meaning sticks with me.

Having decided early on to study medicine, my father knew Greek: at the time of his schooling, and for long afterwards as well, no one could become a physician without studying Greek in high school. This requirement, which today strikes us as baroque, was explained not only by the

Dr. Bernard Weil and Madame Selma Weil with their children André and
Simone in Mayenne (1916)

persistence of tradition, but by the fact that medical jargon is rife with words derived from the Greek. The only one in his family to complete secondary school, my father had done so in Strasbourg, in German perforce, at the *humanistisches Gymnasium*. It was also in Strasbourg that he began his medical studies and performed his military service, most probably in the medical corps, before coming to Paris. He hardly ever spoke of this part of his past; but when he was called up in 1914, he spent just a few weeks close to the front with the ambulance corps before being transferred to the rear by a ruling which affected all those Alsatians whose military service had been with the German army: it was feared that if they were captured they would be treated as deserters.

Whether my father remembered much more from his classical studies than the first line of the Odyssey, which he readily recited, I do not know, but it was because of them that he was later able to copy a number of my sister's manuscripts which were peppered with quotations from the Greek. In any case, I always knew him to be utterly absorbed in the practice of his profession, for as long as he was able: until 1940, his reading was limited to the *Presse Médicale*, to which he was a faithful subscriber. He had neither the time nor, most likely, the inclination to read anything else. An excellent general practitioner, he was held in high esteem by his colleagues for the reliability of his diagnoses. In addition to this talent, his congeniality and openness, along with his irreproachable honesty and true kindness, earned him the adoration of his patients and, during the war, of the personnel in the various hospitals to which he was assigned.

My mother was born in 1879 in the Russian port of Rostov-on-the-Don. Both of her parents belonged to the community of Austrian Jews which constituted one of the finest flowers of nineteenth-century German culture. Her father was Adolphe Reinherz, a prosperous grain merchant who was already well established in Russia before his marriage. He was also a scholar, and my grandmother always treasured the red leather-bound notebook that held his poetry in Hebrew. I often saw this book in her library during her lifetime, but to my chagrin I could not find it again when I looked for it a long time after her death. The entire family left Russia after the pogroms of 1882, settling first in Belgium and eventually – after the death of a beloved son – ending up in Paris, where my maternal grandfather died shortly after my birth. From that time on my Viennese grandmother, with the charming name Hermine, always lived with my family. She was an excellent pianist, and since the period she had spent in Russia she had maintained a lively interest in the Russian language. This interest was what sparked her friendship with Elie Metchnikoff, the renowned biologist, and his wife Olga, a gifted painter who was a pupil of Eugène Carrière as well as the author of a fine biography of her husband. I cannot describe sweet

André and Simone (1911)

Olga Metchnikoff without recalling the way she exclaimed "such lovely blue eyes!" upon being introduced to my wife Eveline after our marriage. Also from Olga, I still have the fine portrait she painted of my grandmother Hermine, whose sad but also peacefully sweet expression follows me around my home in Princeton.

As for Elie Metchnikoff, who died in 1916, the only thing I remember about him is his fine beard; and I am not even sure whether this memory is really of him or of one of the photographs I may have seen of him. But my parents contracted from him a fear of germs which my sister carried to an extreme as a child. It was also Metchnikoff who, hearing of plans to start a class for the children of a number of scientists (including Jean Perrin and Paul Langevin, to name only two) tried to persuade my parents to enroll me in this "school for geniuses." Fortunately for me, these plans fell through.

André and Simone (1911)

My parents were married in Paris in 1905. It was certainly an "arranged marriage," but before long there was a solid base of mutual affection, and theirs remained one of the most stable marriages imaginable. At that time it was virtually impossible for a young physician, however talented he might be, to start out without sufficient funds for setting up an office and purchasing a practice. My mother's dowry was used for this purpose, with the surplus being invested on the advice of an uncle of my mother's, a broker at the Paris Stock Exchange and financial advisor to the family. The investment consisted of Russian and Austrian government bonds, seemingly "sure values" at the time; ten years later, these were to become "non-values," over which my sister and I shared a good laugh. With this one exception, reference was never made to money in our family conversations. After my mother's death, I found several bundles of these bonds in the back of a closet.

My mother was probably just the kind of spouse my father needed. Energetic, passionate even in the slightest of her impulses, and capable of unlimited devotion to her family, around whom she drew a magic circle (which, fortunately, I was able to escape before too long), she was no doubt quick to gain ascendancy over her husband in every matter except his medical practice. He was apparently quite comfortable with this arrangement, until the necessity of leaving Paris in 1940 forced him to abandon his practice and created an irreparable emptiness in his life, a void that was aggravated by my sister's death in 1943. The power to make all decisions about household matters, social life, travel, and vacations thus fell to my mother. She had received an excellent literary and musical education in Belgium, and naturally spoke French and German fluently, and English as well. In Paris, she had apparently been one of the best pupils of Rose Caron, a singer of some reknown at the time, and she continued to cultivate her vocal talent for many years after marrying; and thus it is that I still carry in my head many of the great melodies of Gluck, Mozart, and Schumann – even though I myself have never been able to produce a single note of music.

I learned to read at a rather early age, between four and five, on the upper level (referred to as the "imperial," for reasons I do not know) of the Montrouge-Gare de l'Est streetcar. This distant ancestor of the number 38 bus line led from our house to the Jardin du Luxembourg, where my mother would take us – my sister Simone and me – for walks nearly every day. My mother would make me read the storefront signs along the tram route. I became a voracious reader, devouring everying that fell into my hands. During a vacation at the seaside, my parents were quite amused to find me in the attic, engrossed in Marcel Prévost's *Les demi-vierges*. It was time to see about providing me with an education better suited to my age.

For as long as necessary, my mother supervised our education closely, applying herself to the task with the most intelligent zeal. In 1912, there were no computers to assign each child to a certain school, class, and teacher. After painstaking research, my mother chose an exceptional elementary teacher, Mademoiselle Chaintreuil, who taught the tenth form at the Lycée Montaigne.[1] After several months of tutoring, she deemed me capable, even though I was a little too young, of joining her class at the *lycée*. She was an extremely well-educated person, with whom my mother maintained a long-lasting and very close friendship – closer than she ever enjoyed with anyone else, as far as I know. We called her Tante Gabrielle.

---

1 *Translator's note:* In France, grade levels, or forms, are numbered from the eleventh to the first, the latter being followed by the *"classe terminale"* or final year.

Her favorite reading matter was Amiel's *Journal*. One day my mother, who thought me sufficiently well grounded in reading and writing, expressed her fears regarding "arithmetic" (at that time, for that level, the term "mathematics" was not yet used). Mademoiselle Chaintreuil reassured her: "No matter what I tell him on that subject, he seems to know it already." No doubt she had in mind Plato's theory of remembrance.

"Tante Gabrielle" was also the one who, at the close of the school year, presented me with a copy of Gaston Bonnier's *Simplified Flora*, which I took with me to Ballaigues in Switzerland. Perhaps the ingenious system of classification presented in this book was no less pleasing to me than the Alpine flowers it enabled me to identify; in any case I have a vivid recollection of the meadow-saffron dotting the meadows my sister and I explored there. Even more clearly, when I recall this vacation, I can still see the clouds of little blue butterflies that rose from the tall weeds at our approach, and whose name I was never curious enough to look up.

When school started again, we had moved from the Boulevard de Strasbourg to the Boulevard Saint-Michel, across from the Lycée Saint-Louis, not far from our beloved Jardin du Luxembourg as well as from the Lycée Montaigne. It had been decided at the end of my year in the tenth form that for me to spend a year in the ninth, as I would normally have had to do, would be superfluous: not that my parents sought to "push" me, as is all too often the case, but because the purpose of the ninth form at that time was to reinforce what had been learned during the previous year, and it would not have introduced me to anything new; in that happier era, such a waste of time did not seem desirable. I was therefore promoted directly to the eighth form.

This class was divided into three sections, of which one, made up of the best students from the ninth form, was taught by Monsieur Monbeig. Naturally I was not included in this group. It was my lot to be assigned to an indulgent gentleman whose beard gave him a patriarchal air. As was customary, each class was assigned a weekly *composition*[1] on a topic (French, arithmetic, history, and so forth) that varied from week to week; each week the students were ranked according to their performance on this task. The system was not without its drawbacks, not the least of which was the importance that ambitious parents attributed to these rankings. And yet, is it not strange that competition as a pedagogical motivation has fallen out of favor, whereas the spirit of rivalry has perhaps never been as fierce as it

1 *Translator's note:* The term *composition* was not limited to essays, but might also designate a recitation or the solution of a problem in the context of this weekly exercise.

is now, in virtually all domains? In any case, my mother, learning that the first composition put me at the head of the class, rushed to the principal, telling him: "If my son is ranked first without even having attended the ninth form, it must mean he's in a section that is too easy for him. I want you to transfer him to another class; otherwise, he'll end up doing nothing." The astonished headmaster replied, "Madame, this is the first time a mother has ever complained to me that her son's class rank is too high." But my mother was not one to brook opposition to her wishes, so, against all regulations, I found myself under the paternal rod of Monsieur Monbeig (the father of the geographer Pierre Monbeig, who was to be my friend and colleague in Brazil from 1945 to 1947). He was an exceptional teacher, full of unconventional ideas. For the purposes of grammatical analysis, he had invented a personal system of algebraic notation, perhaps simply to spare both himself and his students time and effort; but it seems to me, looking back, that this early practice with a non-trivial symbolism must have been of great educational value, particularly for a future mathematician. Is it mere coincidence that in India Panini's invention of grammar had preceded that of decimal notation and negative numbers, and that later on, both grammar and algebra reached the unparalleled heights for which the medieval civilization of the Arabic-speaking world is known? At one time it has been thought that young children should be primed for the study of mathematics by being forced to speak of sets, bijections, cardinal numbers, and the empty set. Perhaps I was no less well prepared by my study of grammatical analysis – both verbal and, as it was called at the time, "logical" (that is, propositional) analysis – at the hands of Monsieur Monbeig. I must say in any case that nothing I later came across in the writings of Chomsky and his disciples seemed unfamiliar to me.

Let it not be thought, however, that Monsieur Monbeig taught at a level way over our heads; I do not think that there was anything overly abstract about his teaching. And Saturday afternoons – for during this happy era neither families nor teachers were hounded by the weekend anxiety that obsesses people nowadays (at the time, there was no school on Thursday) – Monsieur Monbeig would read us adventure stories. That year, it was *Le Captaine Corcoran*, and the mighty feats of the tigress Louison kept us on the edge of our seats. On one occasion a "dirty word" appeared on the blackboard. Monsieur Monbeig announced that unless the culprit identified himself (for a punishment that was sure to be minimal, probably no more than a mild scolding), there would be no Louison the following Saturday. Behind me I heard the son of a wealthy family promise a reward of twenty-five *centimes* if his neighbor would offer himself up as scapegoat to save the day. Perhaps no such deal was struck, but my

André and Simone in Mayenne (1916)

confidence in the institution we so comically refer to as justice was shaken once and for all.

In 1914 everything changed: the war irrupted into our vacation on the coast of Normandy. One day in July my father appeared in his reserve officer's uniform, complete with sword. My grandmother thought he was crazy, but shortly thereafter he left for the front. I composed a quatrain, no doubt echoing family conversations:

> Physicians are lucky in war;
> They take care of the wounded and sick.
> They stay behind ambulance doors,
> And the ambulance doors are quite thick.

No doubt my mother and grandmother sought reassurance in such thoughts during those tragic weeks. Soon the first of the wounded began to appear. They were housed in a villa that had been converted into a hospital, where I would go to play checkers with some of the less seriously wounded. I got

hold of a textbook, a geometry text by Emile Borel, which was left to me
(willingly, no doubt) by an older cousin of mine.

At that time, the textbooks used in secondary education in France
were very good ones, products of the "new programs" of 1905. We tend
to forget that the reforms of that period were no less profound, and far
more fruitful, than the gospel (supposedly inspired by Bourbaki) preached
by the reformers of our day. It all began with Hadamard's *Elementary
Geometry* and J. Tannery's *Arithmetic,* but these remarkable works,
theoretically intended for use in the "elementary mathematics" (known
as *"math. élém."*) course during the final year of secondary school, were
suitable only for the teachers and best students: this is especially true of
Hadamard's. In contrast, Emile Borel's textbooks, and later those by
Carlo Bourlet, comprised a complete course of mathematics for the
secondary school level. I no longer recall which one of these fell into
my hands in that summer of 1914, but I still have an algebra text written
by Bourlet for third, second, and first form instruction, which was given
to me in Menton in the spring of 1915. Leafing through it now, I see it
is not without its defects; but it must be said that this is where I derived
my taste for mathematics.

As for my father, after being recalled from the front he spent the
war in a series of military hospitals, except for a period of convalescence
in Menton in 1915, and three months in 1916 in a military column south
of Constantine in Algeria. With the exception of Africa, my mother
wanted to follow him to all these posts, trailing us – my sister and me
– along with her; usually my grandmother Hermine accompanied us as
well. These travels, hardly conducive to a program of regular studies,
were actually of much greater benefit to us than a conventional school
career. On the first of these trips we had to keep our presence secret:
my father had been sent to Neufchâteau, a restricted army zone that was
officially off limits to military families. When we took walks outside we
could hear the thunder of cannons some distance away. This was before
the typhoid vaccine had been developed, and the hospital was full of
typhoid cases. At that time the treatment consisted of cold baths, which
seemed to kill the patients off like flies. It was not long before my father,
overworked and no doubt profoundly dispirited, fell ill himself; he
was sent to Menton for treatment and recovery. This period meant a
delightful vacation for the whole family. Simone and I became avid
collectors of the pretty little shells called "coffee beans." I devised a plan
to surprise my father for his birthday, April 7: my sister was to read the
newspaper to him. I subjected her, I fear, to an intensive regime of study.
Even our walks were devoted to practicing the catechism: "How do you
pronounce *M-A-I-S* ? How do you pronounce *B-E-A-U* ?" But she was

able to read the newspaper to "Biri" (this was our nickname for him) as planned.

From 1914 to 1916, I took only correspondence courses, with the seventh form teacher who would have been mine at the *lycée*. With him I began my study of Latin. As for mathematics, I had for the time being no need of anyone: I was passionately addicted to it. Once when I took a painful fall, my sister Simone could think of nothing for it but to run and fetch my algebra book, to comfort me. To compensate for the lack of regular instruction, I was lucky enough to be given a subscription, starting in the fall of 1915, to the *Journal de Mathématiques Elémentaires* printed by Vuibert. This extremely useful periodical published mainly problems, and principally exam problems, for all levels of secondary education from the third form up. Also published, along with the best solution received by the editors, were the names of those who had submitted correct solutions. I was surprised to discover before long that some of these questions were within my reach. How proud I was to see my name in print for the first time! Soon it was appearing regularly, and then one glorious day, my solution was published. Although the *Revue de Mathématiques Spéciales* is still (or again) in existence, I do not think there is now anything similar to the *Journal de Mathématiques Elémentaires*, and more's the pity.

At the beginning of the school year in October of 1916, my father was either in Africa or on the verge of heading there. We were temporarily at home, on the Boulevard Saint-Michel, and I was admitted to the fifth form at the Lycée Montaigne.

I have often said that a gifted pupil is best off having an excellent teacher every two or three years to provide the impetus he needs, with the rest of the time filled in by a more ordinary instructor. This is about what a student could expect to find in a French *lycée* at the time I was in school. In any case, in the fall of 1916, I was once again lucky enough to meet up with an extraordinary teacher, Monsieur Andraud. He had not only passed the *Agrégation* examination, but he had also earned a doctorate, having written a thesis on Provençal poetry under the direction of Jeanroy.[1] Monsieur Andraud was, he said, among the last students who had to write the complementary thesis in Latin.[2] Jeanroy had even required that he provide a Latin translation of all the Provençal texts cited in the thesis; as

---

1 *Translator's note:* The *Agrégation* is a rigorous, highly competitive examination administered by the French government to prospective secondary school and university teachers.

2 *Translator's note:* At that time a candidate for the *doctorat ès lettres* was required to submit not one but two theses, the "main" thesis in French, and the "complementary" thesis in Latin.

Monsieur Andraud said, that was "a doozy of a Latin exercise." As for the main text of his thesis, he had done his thinking in Latin, something that posed no problem for him. Virtually everything I know of Latin I learned from Monsieur Andraud. At a later date, back in Paris once again, I studied Greek with him, as did my sister. He was qualified for a position at the university, but he always preferred his uneventful fifth-form class at the Lycée Montaigne.

After the productive fall trimester at the *lycée*, I spent the rest of the school year at Chartres, where my father, finally back from Algeria, had just been sent. The class there was so mediocre that I was almost entirely dispensed from attending school, on the pretext of tutoring at home. Once more, we spent part of our summer vacation by the sea. When school reopened in October of 1917, my family was in Laval. I had taught myself enough Greek, and I already knew enough mathematics, to be admitted into the classical section of the third form. I no longer remember whether it was there that I read the first book of the *Iliad*, or whether I read it by myself, for my own pleasure; in any case, that was how I discovered poetry, and that it is untranslatable; for I do not believe this discovery is to be traced back to the precocious enthusiasm I had felt some years earlier for Edmond Rostand, an enthusiasm I had also imparted to my sister. Already we would try to outdo each other in declaiming grand speeches from Corneille or, better yet, Racine. At a recitation *composition* in the third form it fell to me to recite Clytemnestra's tirade to Agamemnon in Racine's *Iphigénie*: this was one of my favorite passages, and I intoned the lines oblivious to my classmates, who were in stitches over my performance. When it was over, the teacher declared, in earnest but not without a touch of humour, that "Weil's declamation was the best," and I was ranked first.

I had at the time, and indeed I still have, Annandale's English dictionary, which includes an introduction to Indo-European linguistics and Grimm's Law as well as fairly detailed etymological information going as far back as Sanskrit. I dreamed of one day being able to read, in the original, the epic poems written in all these languages. My romantic notion of these epics later led me to seek out the advice of Sylvain Lévi.

In 1918 I should have entered the second form, and my literature teacher would have been Emile Sinoir, a *Normalien* and former member of the Ecole d'Athènes.[1] But the war was nearing its end, the Spanish flu

1   *Translator's note:* A *Normalien* was a student at or, as here, an alumnus of, the Ecole Normale, which is one of France's selective "Grandes Ecoles." Admission to these institutions of higher education, which are government-sponsored like the universities but separate from and markedly superior to the latter, is by highly competitive examinations administered nationwide. The Ecole Normale is attended

was on the rampage, and my family was preparing to return to Paris. Instead of going back to the *lycée* in the fall, I was tutored privately by Mr. Sinoir. A humanist of great distinction, he made a lasting impression on me; I continued to correspond with him for quite some time after he taught me. I read Plato with him, and he would assign essays for me to write. Once I had to write up a character in the style of La Bruyère; another assignment was to write "a letter to a stationmaster inquiring after a lost object." As can be seen from such examples, instruction was not as narrowly bookish at that time as it is now supposed to have been.

Back in Paris in our apartment on the Boulevard Saint-Michel, I was entrusted to Mr. Andraud for Greek lessons. These took place in our dining room, where there was also a cage with two canaries that my sister doted on. Mr. Andraud's warm meridional voice almost never failed to elicit imitations by the male canary, which would launch into endless trills as soon as it heard his Toulouse accent.

It was understood that in October I was to enter "C" track (Latin and Sciences) of the first form at the Lycée Saint-Louis. My parents quite correctly believed that there were still serious gaps in my mathematical knowledge. They contacted Mr. Collin, who was to be my teacher in the first form and whom I ended up having again the following year, in the *classe terminale*. He was the one, then, who prepared me to enter the first form.

It would be impossible for me not to expand a little on the virtues of this remarkable teacher. I know little about his personal life, except that he was a bachelor and lived in a small apartment near the Botanical Garden, and that his favorite summer activity was bicycling in Auvergne. I never noticed whether his intellectual horizons extended beyond the level of mathematics that he was responsible for teaching; perhaps he was too reserved to let on. Like the great majority of *lycée* teachers I have known, Mr. Collin had no problems with "discipline," in the vulgar sense in which the word is used today. Once, when I was no longer his student, he told me that the first class he had been assigned to (a preparatory class to Saint-Cyr, if I recall correctly) was known as a tough bunch of trouble-makers.[1] From the very first moment, he realized that the students had planned to put him to the test. He sat down in front of

by the most promising candidates preparing for the *Agrégation* examination for positions in secondary and higher education. The Ecole d'Athènes is an institution maintained by the French government in Athens for budding archaeologists.

1   *Translator's note:* Saint-Cyr was a military school, admission to which was by competitive examination; the preparatory class trained students intending to take the examination.

the class and without uttering a word stared at the students for the entire hour. Never again did he have the slightest trouble with them. I do not think any teacher could have been better than Monsieur Collin at developing both rigorous thinking and creative imagination in students. He would send a student to the blackboard and put forth a problem. If it was a geometry problem, he would make sure the figure was correctly drawn. Often ten minutes would pass in total silence. Not only the student at the blackboard, but all the others at their desks would try fervently to find the solution: in Monsieur Collin's class, even the "professional" dunces – those who were proud to be so – were anxious to prove that they were no duller than the rest of us. After a reasonable amount of time, Monsieur Collin would ask: "Who's got it?"; a few hands would go up. When a sufficient number of hands had been raised, the class worked out the solution as a group. If everyone was stuck, Monsieur Collin would say: "I shall make some remarks," and his hints would put us on the right track. On the other hand, definitions had to be memorized, and Mr. Collin was merciless towards any gap in solutions or proofs. With him, mathematics was truly a discipline, in the fullest sense of this beautiful word.

Naturally the oral questions and problems were but indispensable complements to the formal lectures that constituted the core of his instruction. Except for definitions, Monsieur Collin did not dictate a text to us. It was understood that, as *lycée* students, we were supposed to be learning the art of intelligent note-taking; and indeed, I believe there is no better training for the mind. Everyone took notes, and in cases of doubt compared them with those of his classmates. Although a number of very good textbooks existed, no teacher worth his salt – in mathematics or any other subject – was content with these alone.

It has become fashionable to criticize the mathematics curriculum for first form and "*math. élém.*" levels, but even the most questionable, and most questioned, aspects of these programs had their good points. The geometry of the triangle and the focal theory of conics lent themselves to problems that sharpened the geometric imagination; problems on geometric loci, even the so-called "trinomitis," accustomed us to "complete enumerations" in the sense of the Cartesian method. In truth this instruction, so disparaged today, left me with favorable memories.

But I am getting ahead of myself. What I remember most about Monsieur Collin's lessons prior to entering the first form is that he showed me once and for all that mathematics operates by means of rigorously defined concepts. I had not imagined that a function could be expressed otherwise than in an algebraic formula; in the books that had come into my possession I had noticed nothing to dissuade me of this naive idea. I do not recall in what terms Monsieur Collin taught me the definition of the word "function." He could not have used the language of Bourbaki, which was as yet nonexistent, or even that of set theory, with which he was probably barely acquainted. What is important is that once the definition was given, he would not tolerate anyone's using the word "function" for anything not corresponding to this definition. If a function were given by a certain formula in one interval, and elsewhere by a different one, it was still a function, since the definition said so.

I think there is no one, with the sole exception of Hadamard, from whom I learned more about mathematics than from Monsieur Collin. Before I became his pupil, I was basically self-taught; he made a mathematician of me, and he did so above all by means of his unrelenting criticism. His intransigence was all the more effective as, in order to enliven the weekly exercises (which I found somewhat too easy), I had set myself a limit of two pages into which everything had to fit. There was a strong temptation to take short-cuts, saying "it is obvious that..."; Monsieur Collin taught me never to use this word. "If it were obvious," he said, "you would not feel the need to say so; if you say so, that means it is not obvious." He is the one who taught me how to write up mathematics.

Though the Lycée Saint-Louis claimed, not without reason, to be the best scientific *lycée* in France, it must not be thought that the humanities were neglected there. I already knew enough English, and I had a sufficiently developed taste for poetry, that I had little to learn from an English teacher who was intelligent but perhaps too didactic. On the other hand, the history teacher successfully communicated to me just how intellectually stimulating this discipline could be. Bearded and broad-shouldered, he had a rare gift enabling him at will to throw the class into an uproar with some provocative statement, and to quiet it instantaneously. He encouraged us to go beyond our textbooks. These were not my first delvings into history: as early as 1916 I had been an avid reader of Fustel de Coulanges' *La Cité Antique*; in Laval I had also read Augustin Thierry's Merovingian chronicles and a history of Greece (by Petit de Julleville) which was essentially, if memory serves me, an adaptation of Thucydides. My first form teacher suggested that I go to the Sainte-Geneviève library, which was near both school and home, to read Mathiez on the French Revolution.

In French literature and in Latin, our professor was the brilliant Charles-Brun, a militant meridional regionalist and a somewhat bohemian man of letters who was the first to make fun of his offbeat appearance.

"Those who are fond of me," he would say, "refer to my hair as 'his wind-tossed locks, which shun the comb'; the rest say 'his filthy unkempt mange.' " I would go to his house on the Rue Delambre to read Greek, for Greek was not included in the first form curriculum of the scientific track at the *lycée*. We read Plato's *Crito*; his teaching enabled me to read the *Phaedo* and the *Pro Corona* during the summer vacation. Charles-Brun was not overly conventional. Nevertheless, it was thought then that it was best to learn to observe the rules before allowing oneself to break them; every essay had to have a plan, a logical development, and a conclusion. One day the topic was a statement by Renan enumerating the qualities that make up a nation; we were to apply the list to France. In 1920, this was certainly a subject that could crop up on the baccalaureate examination. I took pains to demonstrate, point by point, that France had none of the necessary qualities, and I concluded: "this explains why regionalism continues to make progress under the auspices of Charles-Brun." The assignment was returned to me with a grade of 12 out of 20[1] and the comment: "This is rather amusing, though too long... and you will fail the baccalaureate exam."

In fact, there was no great danger of my failing – although my success was jeopardized for other reasons. I was significantly younger than the minimum age required to take the examination; my request for special consideration was denied. Since I could not start "*math. élém.*" without this first diploma, alternative plans were discussed. For a year I had been attending a carpentry course every Thursday in a primary school on the Rue Cambon, and I had taken quite a liking to it. It was not yet obvious, either to my family or to my teachers, or even to me, that I was destined for a career in mathematics. If circumstances later forced me to turn to the physical sciences, or to engineering, a little practical experience could come in handy; it was suggested that I spend a year in a vocational or similar school studying metalworking or applied mechanics or electricity. Though I had not exactly been thrilled with the thoroughly pedestrian instruction in physics and chemistry I had received in the *lycée,* I looked upon this necessary interlude without distaste. In any case, as chance had it, I was spared. The school inspectors paid their annual visit during the third trimester. Monsieur Collin called me to the blackboard, took pleasure in making me shine, and then explained the situation to the inspector, who was influential in the Ministry of Education; my renewed request for a dispensation was granted. During the summer, Monsieur Collin was transferred to a class of "*math. élém.*," where I found him in the fall of 1920.

---

1   *Translator's note:* In the French system of grading (based on a total of 20 points), a decent grade.

At that time, good students were encouraged to take baccalaureate exams in both "*math. élém.*" and "philosophy": the double baccalaureate gave an advantage to those applying for admission to the Ecole Polytechnique.[1] I followed this plan. As far as my school records went, it was a success; but despite the efforts of a worthy teacher, philosophy never did "take" with me: this discipline appears to be incompatible with my way of thinking. On the examination, I was given a question relating to Kant and Durkheim. I was familiar enough with the textbook in use to write an acceptable essay, but in truth I had never read a sentence written by either of these authors, and I was shocked to receive a grade far superior to what I thought I deserved. When the philosophy examiner inquired as to my plans for the following year, I replied without hesitation: "I will be studying for the entrance exams for the Ecole Normale." "In philosophy, of course?" "Oh no, sir; in mathematics." It seemed to me that a subject in which one could do so well while barely knowing what one was talking about was hardly worthy of respect. Youth is unjust.

It is true that, in the meantime, I had become acquainted with Grévy, and then with Hadamard, and my future plans were taking shape. At Saint-Louis there were four "*taupe*" sections: this is the name commonly given to courses preparing students for the entrance examinations of the Ecole Polytechnique and the scientific section of the Ecole Normale. There was also an intermediate class, known as "*hypotaupe*": here the curriculum was the same as for the *taupe* sections, minus the requirement to take the examinations at the end of the year. Of the four *taupe* professors, Grévy and Michel were considered the best. Monsieur Collin introduced me to Grévy, who trusted Collin's judgment implicitly; both of them thought me capable of joining Grévy's class with a view to taking the Ecole Normale examination as soon as the following school year. Even more importantly, Grévy introduced me to Hadamard, who had been his classmate at the Ecole Normale.

All those who were acquainted with Hadamard know that until the end of his very long life, he retained an extraordinary freshness of mind and character: in many respects, his reactions remained those of a fourteen-year-old boy. His kindness knew no bounds. The warmth with which Hadamard received me in 1921 eliminated all distance between us. He seemed to me more like a peer, infinitely more knowledgeable but hardly any older; he needed no effort at all to make himself accessible to me. Soon he had the opportunity to do me a favor that had a decisive effect on my

---

1   *Translator's note:* The Ecole Polytechnique is one of the best-known of France's prestigious "Grandes Ecoles."

future. Every year the Lycée Saint-Louis awarded an endowment prize to the best student in "*math. élém.*" The prize consisted of the equivalent in books of the annual interest on the endowment fund. I was allowed to choose these books myself, and sought Hadamard's advice on the selection. Thus it was that at the annual "solemn award ceremony" (anyone who attended a French *lycée* at the time can picture the scene) I received Jordan's three-volume *Cours d'Analyse* and Thomson and Tait's two-volume *Treatise of Natural Philosophy*. Thanks to Hadamard, I learned analysis from Jordan (infinitely better than learning it from Goursat, as most of my classmates did) and I was initiated into differential geometry by Thomson and Tait. Naturally I did not read these hefty volumes immediately; I think, however, that I did begin with Jordan the following year.

In 1921, relativity – or, as it was called at the time, "Einstein's theory" – was all the rage. The topic was too widely discussed, even in the newspapers, for me not to wish to become familiar with it. I did so by reading Eddington. As a kind of formal manipulation, tensor calculus seemed easy to me, although naturally I was unable to appreciate its geometric aspect. During a vacation in the Black Forest, I got it into my head to use my parents as guinea-pigs for a series of explanations of relativity in the course of our family outings. They very patiently complied. The next year, when I was in Grévy's preparatory class, Einstein was invited to give a series of lectures at the Collège de France.[1] Admission was by ticket only: I received mine thanks to Hadamard, I suppose. Everyone who moved in scientific or philosophical circles or in high society came, in such droves that the Garde Républicaine had to be called in to control the crowds. With such an atmosphere, a high-level scientific exchange was hardly to be expected; if there were serious scientific discussions on this occasion, as I am sure there must have been, they must have been limited to the fit and the few – Einstein, Langevin, and a handful of others. Paul Painlevé, who felt it his duty to take part in the public discussions, attended faithfully, adding to their picturesque quality, but still I felt slightly disappointed. On the other hand, I witnessed a memorable discussion between Elie Cartan and Einstein, at the home of the philosopher Xavier Léon. Einstein had based his general theory of relativity on a foundation of classical Riemannian geometry, which, to use Cartan's terminology, is a geometry with curvature but without torsion; no doubt Einstein was

---

1   *Translator's note:* Known under various names since its founding in 1529, the Collège de France is an educational institution with approximately fifty chairs in all subjects. All lectures at the Collège are open to the public; the Collège neither administers examinations nor grants degrees.

completely unaware at that time that other types of geometry could be envisaged, whereas Cartan's perspective made it possible to go much further. At Xavier Léon's, Cartan pointed out to Einstein that it was also possible, for example, to conceive of geometries with torsion but without curvature (in fact, this is the type of geometry with absolute parallelism that Einstein himself later introduced). Naturally, I did not understand the meaning, much less the implications, of what Cartan said; still, it made enough of an impression on me that it came back to me years later, when Cartan's ideas had become familiar to me.

In the *taupe*, of course, the student acquires – or at any rate acquired at that time – a facility with algebraic manipulation, something a serious mathematician is hard put to do without, whatever some might say to the contrary. As for Grévy's own course, there is little to say about it now. It was, I believe, a serious, honest course, well-planned, and doubtless superior to the texts available in bookstores; in any case, it rendered them superfluous. As far as set programs go, it would have been nearly impossible to produce a better one. With respect to method and to preparation for the competitive examinations, this instruction unquestionably achieved its goal – but I do not believe it really taught me very much. A year of *taupe* was salutary; a student could, without harm, and sometimes even fruitfully, tolerate two years of it, counting a year of *hypotaupe*. But the veteran *taupins* who had failed – sometimes through no fault of their own – and spent three, or even four, years there, were a rather frightening bunch: theirs was not an enviable fate. Luckily, I was not to share it.

Fortunately, too, the preparatory course did not take up all of my time that year. I began to study Jordan. Also, my precocious and romantic attraction to Sanskrit gave one of my father's friends the idea of introducing me to the leading scholar in the field of Indian studies, Sylvain Lévi. At that time, university professors worked and received visitors at home, amongst their books. Neither Hadamard nor Sylvain Lévi, nor any of their colleagues I believe, had an office at the Collège de France; only in the science laboratories was a nook for the professor sometimes to be found. When these scholars chose not to receive a visitor at home, they would usually arrange to meet him in the antechambers of the Academy on a day when it was in session.

When Sylvain Lévi received me at his home in the Rue Guy-de-la-Brosse, he said to me: "There are three reasons for studying Sanskrit," and he enumerated them: I believe they were the Veda, grammar, and Buddhism. "Which of these is yours?" I didn't dare tell him that I was impelled by none of the three, but simply by the naive notion I had of Indian epic poetry. I had already learned – where, I do not recall

– the alphabet and one or two declensions, and I asked his advice on how to proceed. He told me that the best textboook by far was Bergaigne's, but it was out of print; otherwise I should obtain a copy of Victor Henry's text. This I bought immediately, and I lost no time in putting it to use.

It was also during my final years in high school that my enthusiasm for literature, and above all for Greek poetry, was enhanced by a taste for old editions of the Greek and Latin authors. Some of these were still to be found along the quays of the Seine. Reading these old editions made me feel even closer to the writers of antiquity – not that their contemporaries themselves had read such volumes, but the sixteenth-century editions upon which my predilection soon settled were the work of the great humanists such as Estienne, Aldus, and Budé; and they were not so very far removed from the manuscripts. I knew that the modern editions might boast more reliable and more accurate texts; but that was not important to me. Moreover, I came to see that the more recent editions infallibly suffer from an excess of punctuation. These crutches, while helpful at times, are not in the spirit of the text, and sometimes it is easier to understand Plato or Demosthenes by yielding to the rhythm of a long period, which may better express what the author had in mind to convey to his reader, than by using a text that has been chopped up with periods and commas. Besides, anyone who loves fine typography cannot fail to admire these masterpieces produced in the early days of printing and throughout the first century of its existence. When I was a student it was already extremely difficult to find the great incunabula which are the glory of the presses of Rome, Florence, and Venice; and when they did turn up, they were well beyond the bounds of my modest budget. But among the booksellers in the Latin Quarter, sometimes even in the boxes along the Seine, it was still possible to find – at prices that even I could afford – beautiful books printed by Aldus, Estienne, or Simon de Colines. I fell in love with them at first sight. In a tiny shop beneath the galleries of the Palais-Royal, I made the acquaintance of a learned old bookseller, who took a liking to me. This man, whom I called "Père Gauvain," taught me how to recognize the great printers' marks and how to read the ligatures which, faithfully copied from manuscripts, used to be employed in printed Greek texts, and which to a large extent account for their beauty. Selling books could not have been farther from Père Gauvain's mind; yet it was from him that I purchased my Greek and Latin edition of the *Iliad*, a 1560 *sextodecimo* which was my inseparable companion for many years. A little later he alerted me to a copy of Aldus's *Plato* (the *princeps* edition of 1513), which one of his fellow booksellers, who had been unsuccessfully trying to sell it for some time, was willing to let go for a song – relatively speaking. This time, my parents

A.W. on vacation in Baden-Baden (1921)

had to come to my aid. Père Gauvain also taught me how to use Brunet's admirable *Manuel du Libraire*. Luckily this congenially erudite master-piece had just been reprinted in a collotype edition in Germany. Using it, I was able to set up an imaginary library à la Malraux, in which each Greek or Latin author I had chosen was represented by a few of the finest old editions. All that remained for me to do was to seek out the actual books. By dint of great patience, I managed quite inexpensively to put together a rather nice collection. But around 1930, many American libraries thought to assume the trappings of nobility by building collections of "rare and precious" books; the prices went sky-high, and were soon way beyond my means.

But in 1922 I couldn't see that far ahead. I said my goodbyes to the *lycée,* and was off to the Ecole Normale.

Chapter II
At the *Ecole Normale*

At the "Ecole," as we used to call it, the students were divided into groups sharing studying quarters. My first concern, even before school started, was to find companionable study-mates. There were five of us: besides myself, Labérenne, Delsarte, Yves Rocard, and Barbotte (who, having placed first in the entrance examination, was our "cacique"). Labérenne, who had been in Grévy's preparatory class with me, was a tall young man with an open mind; he was a good companion, not at all an egghead. Delsarte came from Rouen, after only one year in the *taupe*, like me. Rocard, from the Lycée Louis-le-Grand, lost no time in filling his locker with large black notebooks already covered with the tiny but quite legible handwriting in which he recorded his own ideas and calculations on the kinetic theory of gases. None of us four was inclined to docility. The same could not be said of our cacique, who had been in the *taupe* in Versailles: the son of a military man, he had respect for authority and was little disposed to practical jokes; the administration always thought highly of him. This was not the picture I had been given of him by some of his classmates in Versailles. He ended up feeling rather uncomfortable in our company: we must have seemed dangerously heretical in his eyes.

After finding a study group, my second concern was the library. The humanities library, presided over by the renowned Lucien Herr, was extremely accessible to students. In contrast, the sciences library, which was reached through a room almost entirely taken up by a megatherium skeleton (we called it the "mega"), was open to students only one or two hours a week. The staff member in charge, Marcel Légaut, who later went on to write a number of devotional works, had managed to get his duties reduced to the barest minimum, on the pretext of writing his doctoral thesis. My first visit to Vessiot, our estimable scientific director, was to tell him that this minimum was not sufficient. He responded with the wisdom of Solomon, creating especially for me the position of library assistant (non-remunerated, of course) which gave me the key to the library in exchange for nominal custodial duties. Soon I was in the habit of spending long stretches of time in the library, both by day and, sometimes, by night.

Primarily as a result of exchange arrangements, the library was well equipped with periodicals from all over the world; its other holdings were fairly good as well. I do not know how there happened to be a complete

A.W. playing chess during vacation in Belgium (1922)

collection of the newspaper *Le Temps* from the war years.[1] I had long since
given up the flag-waving sentiments I had acquired in childhood from
reading Paul Déroulède and the newspapers. Besides, while my father,
good Alsatian that he was, was still an old-fashioned patriot, chauvinism
had not been considered acceptable in my family. My father himself, in his
youth, had apparently entertained anarchist sympathies; he had surely been
pro-Dreyfus, and was still in some vague sense a *radical-socialist*.[2] In any
case we hardly ever spoke of politics in my family, except to comment on
the latest news. Moreover, a younger sister of my mother's had married in
Germany and was living in Frankfurt. Despite wartime laws prohibiting all

---

1   *Translator's note:* The news daily *Le Temps*, notable for its international scope,
    played a key role in the political life of the Third Republic. It was resurrected after
    World War II as *Le Monde*.
2   *Translators note:* The term *radical-socialiste* designated a political party that was
    somewhat left of center, but neither "radical" nor "socialist" as these terms are used
    in English.

A.W. and Simone reading among the pine-trees (summer, 1922)

communication with "the enemy," my grandmother Reinherz had never stopped receiving news via Switzerland. The word *"Boche"* ("Kraut") was not part of our vocabulary. Even so, I was not displeased to get my hands on the collection of *Le Temps*, so I could read first-hand the "eyewash" palmed off on the French public during the war (this was before the term "brainwashing" was in vogue). I confess that I was somewhat surprised to find articles by Debussy explaining in all seriousness why German music did not deserve its reputation, and articles by Emile Picard saying more or less the same thing about German mathematics. I set about finding the famous manifesto which for so long fueled propaganda against the "Teutonic barbarians." This text, which had appeared in 1914, was signed by ninety-three German intellectuals in an effort to counteract the disastrous effect that the bombing of the Reims Cathedral had had on world opinion. The text was always cited by the Allies as a monstrous attempt to rationalize the bombing. I found its tone to be one of moderation; it fittingly deplored the damage to the cathedral, and only one statement it contained could really be considered shocking: this to the effect that it would not have been right to risk the lives of German soldiers even to save a Cathedral. This idea did indeed seem barbaric to me. Not that one cannot argue, abstractly speaking, that a human life is worth more than a cathedral; this is a metaphysical thesis which – though I do not accept it – can be defended. To my mind, what was barbaric about this statement was that it specified that the lives to be preserved were those of German soldiers, and not of others. How could I suspect that this same argument, with "our boys" instead of "German soldiers," would one day become the official doctrine of the defenders of civilization, and would go on to be multiplied *ad infinitum* the world over, with similar substitutions, spawning the most grotesque caricatures imaginable?

To conclude with the "manifesto," I will note here what I learned much later, in Germany. First of all, Hilbert, who always behaved with the utmost dignity throughout the war, refused to sign it – though I do not think that his name was familiar enough to me in 1922 for me to notice that it was absent from the list of signatories. Second, I have been told that many of those who signed, including Felix Klein, had not seen the text; they had simply been asked over the telephone to support what was put forth as a patriotic duty. Only those who have no inkling of how petitions, protests, and declarations of all sorts are peddled among the intelligentsia would find this surprising.

And so, back to my *bibli* (this was our way of referring to the library at the Ecole): it happened that, when I entered the Ecole, I had a fairly good practical command of German. I had never studied the language, but my parents, who both spoke it fluently, were accustomed to using it when my sister and I were not supposed to understand what they were talking about – a highly effective method of language instruction! As a result I was able to make use of German books and periodicals without much difficulty. In addition, reading Jordan had given me an advantage over the others in my class. I therefore excused myself from attending Goursat's course, which was generally considered compulsory for new "conscripts." I have not had occasion to regret this decision. Indeed, we were given a very free rein, limited only by the knowledge that failing the *Licence*[1] examinations at the end of the year would mean losing one's place at the school and, of course, all the attendant privileges. That same year I began to attend and even to take part in Hadamard's seminar at the Collège de France. This seminar was truly one-of-a-kind, and would still be so today if it were being held.

The word "seminar" has become rather cheapened by overuse: seminars are a dime a dozen nowadays. A history of seminars would have

---

1   *Translator's note:* The *licence* is, loosely speaking, the equivalent of the Anglo-American Bachelor of Arts degree.

to go back at least as far as Jacobi.[1] In Paris, while I was at the Ecole and for a long time afterwards, there was only one seminar worthy of the name, and this was Hadamard's. At the beginning of the year, we met in the library of his home on the Rue Jean-Dolent, where he handed out mathematical papers to be reported on. These papers were for the most part the offprints which he had received from all over the world – or at least those which appeared to him worthy of discussion. To these he added titles of various provenance, as well as titles proposed by others of us, for he was quite open to suggestions. Most of these works had been published in the last two or three years, but this condition was not hard and fast. As to the subjects covered, his aim was to provide as extensive a panorama of contemporary mathematics as possible. If it was not exhaustive, this was at least his goal. For every title he announced, he would ask for a volunteer, often explaining briefly why the paper excited his curiosity. Once the titles had been distributed, dates were set for reports to the group, and after some general chit-chat, we all left.

At that time the seminar took place once a week; later it met twice a week. Among its participants were highly accomplished mathematicians as well as beginners. Paul Lévy, who had been Hadamard's pupil, was among the faithful. Hadamard behaved as if the exposés were primarily for the purpose of informing him personally; it was to him that we addressed ourselves and especially for him that we spoke. He understood everything, as long as it was explained well; when the summary was not clear, he would request clarification, or else – not infrequently – supply it himself. He always reserved the option of adding his own remarks at the end, sometimes in a few words and at times in a more leisurely fashion. Never did he appear conscious of his own superiority: whoever was giving the exposé (it is on purpose that I do not use the word "lecture," for in front of Hadamard it was impossible for it to come across as a lecture) was treated as an equal. This was true even for me, callow student that I was when, shortly after entering the Ecole, I was accepted as a participant – no mean mark of favor. I believed I already had some ideas on functions of several complex variables; I had made several observations (which I thought to be original, and which may perhaps have been so) on the domain of convergence of power series in several variables, generalizing Hadamard's classic theorem on series in one variable; but even more importantly, I had just discovered Hartogs' works in the Ecole library. Although they had already been around for some time, they were little known in France, and had never

1 *Translator's note:* Carl Gustav Jacobi (1804-1851), one of the greatest mathematicians of the nineteenth century.

been the subject of an exposé in Hadamard's seminar. I proposed this topic, and he received my suggestion with pleasure.

The *bibli* and Hadamard's seminar, that year and the following ones, are what made a mathematician out of me. I did attend other courses: Picard's at the Sorbonne, and Lebesgue's at the Collège de France. These two courses, given by strong personalities of two very different types, were both instructive – but I have attended too many classes and lectures in my life to think it worthwhile to describe them all. Every Tuesday at five o'clock in the afternoon, entering Lebesgue's classroom, we looked admiringly upon the ancient Foucart who, having covered the blackboard with Greek inscriptions, was coming out of the same room, supported by two of his faithful pupils. He was the last survivor of the era when professors at the Collège de France were appointed for life, and he was not much under a century old; his two students did not appear to be much younger. In winter, in this same room, there would also be neighborhood drunks who had come in to warm up during the lectures, which were open to the public without exception. They had no compunctions about dozing off, but they were not indifferent to the sounds by which they were lulled into sleep – at least we observed that they did not stay long in Lebesgue's lectures, and we would make bets on how long they would last. I don't think any of them held out for more than eight minutes.

That same year, I began to read Riemann. Some time earlier, and first of all in reading Greek poets, I had become convinced that what really counts in the history of humanity are the truly great minds, and that the only way to get to know these minds was through direct contact with their works. I have since learned to modify this judgment quite a bit, though I have never really let go of it completely. My sister, however, who had come to a similar viewpoint – either on her own or perhaps partly under my influence – held on to it until the very end of her too short life. During my year of instruction in philosophy, I had also been struck by a phrase of Poincaré's which expresses no less extreme a position: "The value of civilizations lies only in their sciences and arts." With such ideas in my mind, I had no choice but to dive headlong into the works of the great mathematicians of the past, as soon as they were materially and intellectually within my grasp. Riemann was the first; I read his inaugural dissertation and his major work on Abelian functions. Starting out thus was a stroke of luck for which I have always been grateful. These are not hard to read, as long as one realizes that every word is loaded with meaning: there is perhaps no other mathematician whose writing matches Riemann's for density. Jordan's second volume was good preparation for studying Riemann. Moreover, the *bibli* had a good collection of Felix Klein's mimeographed lecture notes, a large part of which is simply a rather discursive, but intelligent, commentary fleshing out the extreme concision of Riemann's work.

Through all of this, I took pains not to forget Sanskrit. Jules Bloch was at that time teaching a beginners' course at the Sorbonne. This was where I met the *archicube* Georges Dumézil, destined for a brilliant career.[1] In addition to meticulous instruction in the rudiments of the language, Jules Bloch was constantly serving up his own sparkling observations on Indian languages and civilization and a host of other subjects. At the end of the year, wishing to devote part of my vacation to reading a Sanskrit text, I went to Sylvain Lévi for advice. From a shelf in his library, he pulled a small volume bound in red velvet. It was a "native" (to use the term in vogue at the time) edition of the *Bhagavad Gita*. "Read this," he told me. "First of all, you cannot understand anything about India if you haven't read it" – here he paused, and his face lit up – "and besides," he added, "it is beautiful." With the help of a dictionary lent to me by Jules Bloch, and of the English translation of *Sacred Books of the East*, borrowed from the Ecole library, I read the *Gita* from cover to cover in Chevreuse, at a place known as "La Guinguette," in the big overgrown garden belonging to a small house that my parents had built a year or two earlier. I later found out that Sylvain Lévi pitied anyone who chose to see this poem as a later addition to the *Mahabharata*: not that the text of this immense epic has not been reworked and added to throughout the centuries; but, Sylvain Lévi wondered, how could anyone not see that the *Gita* is the heart of the work, and that the rest must necessarily have been composed around it? One might as well say that the pit was added to the fruit as an afterthought.

The beauty of the poem affected me instantly, from the very first line. As for the thought that inspired it, I felt I found in it the only form of religious thought that could satisfy my mind. My sister and I had been brought up without any semblance of religious education or religious observance. Nothing was left in our family to recall our Jewish background in any way, except for my paternal grandmother, a good old Alsatian woman who spoke dialect more readily than French and who remained faithful to tradition. She was very fond of us, and called me *mon hammele* ("my little lamb") when I was a young boy. I remember her pure silver hair, so lovely that one day a hairdresser stopped her in the street and offered to buy it from her. She lived with my Uncle Oscar, my father's younger

1 *Translator's note:* The term *"archicube"* is slang for former students of the Ecole Normale. Georges Dumézil (1898–1986), a French linguist, historian, and authority on Indo-European mythology, is best known for his conception of the tripartite structure operative in religious beliefs and social organization; he held a chair at the Collège de France and was named to the Académie Française.

brother; my sister and I saw her seldom, and made no attempt to explain to ourselves what appeared to us to be culinary fetishes. It is difficult to believe now, but I am sure I did not know I was Jewish until I was ten or twelve years old, and when I found out, I certainly did not consider it of any importance. By "Jewish," I mean of Jewish descent, according to the traditional definition of these words; these days everyone seems to define Jewishness as he sees fit, and the Archbishop of Paris can apparently declare himself Jewish without worrrying whether or not the Chief Rabbi concurs. If I had asked myself as a child what Judaism meant, I would no doubt have concluded that this concept, along with the complementary concept of anti-Semitism, had to do with history or anthropology, and was no concern of mine. In any case, I would not have associated it with the idea of religion. In truth, the Dreyfus affair had not occurred so very long before; my parents must have remembered it vividly, but they never talked about it, and for a long time I knew nothing of it.

In the *lycée*, on the other hand, I had read a fair amount of Pascal, first the *Provinciales* and then the *Pensées*. I had admired not only the beauty of his writing, but the coherence of the system forged by this powerful mind. Still, between admiring this system of belief and being tempted to adhere to it, there was an unbridgeable gulf. With the *Gita*, however, I felt as if I could enter into its world wholeheartedly. Later on, I felt at home in India thanks to the *Gita*; and much later, it was also thanks to the *Gita* that my sister's way of thought did not seem at all foreign to me.

My second year at the Ecole was just a continuation of the first, with an even greater degree of freedom since I had blasted through all my examinations in one year, and my time was now my own. I must have read Fermat that second year. Our director was giving a course at the Sorbonne on Lie groups. How much better off we would have been if Cartan had given it! But the chair in transformation theory had been created for Vessiot, and Cartan was too scrupulous to encroach upon a colleague's territory. This course was considered more or less compulsory for students in my class. I went once, was bored stiff, and never returned. I did regularly attend Jules Bloch's course on the *Veda* at the Ecole de Hautes Etudes. Every week in the stairway there I would cross paths with Vessiot, who was going upstairs to teach his course as I was leaving Jules Bloch's. I always greeted him politely. This did not improve my standing with him.

I also attended Meillet's lectures on Indo-European linguistics and Sylvain Lévi's on Kalidasa's *Meghaduta*. These two teachers were without peer. Meillet, already nearly blind – this was said to be from deciphering numerous undecipherable texts from Central Asia – would hold forth extemporaneously, treating his audience to observations which even my ignorant mind could tell were strikingly original. Sylvain Lévi read and explained the Kalidasa poem, verse by verse, in his slightly muffled voice. First he would read the Sanskrit text: for him, Sanskrit was a living language, and the pandits in India have long preserved the memory of the speeches in impeccable Sanskrit which he could deliver impromptu whenever the occasion called for it. After the Sanskrit came the Tibetan translation, of which some of those present, including me of course, understood not a word. Then came Lévi's commentary, and finally he would offer a French translation, always both beautiful and precise. This poem consists principally of a long speech made by a Yaksha in love to the "messenger cloud" (the *meghaduta* of the title) which will relay it to his distant best-beloved. Sylvain Lévi excelled in bringing out the marvels of this poem; I can still hear him almost whispering the beginning of the tenth stanza:

> *mandam mandam nudati pavanah...*

and in French:

> *doucement doucement te pousse le vent*
> (gently gently blows the wind upon you).

The third year at the Ecole was generally devoted to preparing for the *Agrégation* examination. My work showed the effects of this exigency – not that I spent a great deal of effort studying for the examination, which I was required to sit at the end of the year; but I felt guilty doing anything else. Thus I acquired a taste for music, and I attended concerts frequently. For five or six years, my family had made me study the violin. My grandmother Reinherz, passionately devoted to music, was herself an excellent pianist. Long before, while living in Belgium, she had lost her dearly beloved son, who was both a brilliant student of the law and an artist. He had been the prize student of his violin teacher, who had bequeathed him a Guadagnini, comparable to a Stradivarius, I believe, in the eyes of connoisseurs. My grandmother had saved the instrument reverently, cherishing the hope that some day one of her grandsons would show himself worthy of it. Occasionally it was entrusted to me. It had a marvelous sound, but alas, I knew I was not equal to the challenge. While one can play the piano simply for pleasure, even without talent, the violin requires at least some gift, which was cruelly lacking in me. My efforts to play it revealed to me both the beauty of music and the hopelessness of trying to recreate this beauty myself. I resigned myself simply to listening and subscribed to the weekly *La Semaine Musicale* (now defunct), listing not only the programs for Parisian concerts, with short but substantive analyses of the pieces played, but also the programs of organ music for

Sunday church services in Paris. It was thus that I heard Bach's *Passacaglia* for the first time on the organ at Saint-Jacques du Haut Pas. A whole new world thus opened up to me.

The *Agrégation* examination consisted of four problem sessions and two lectures to be delivered in the presence of a jury, on selected topics from the secondary school curriculum. Preparation for the problems meant reviewing those from the preceding years, most of which were rather cockeyed. I did a modicum of such work, and had no occasion to regret it. On the other hand, the classroom sessions are an extremely useful exercise, particularly if followed by a thoughtful critique. This exercise teaches one how to plan an exposé; one also learns – at any rate, one ought to learn – how to conduct oneself appropriately at the blackboard: to speak to the audience rather than to the board, and to emphasize the points one deems important. As a student at the Ecole, and then later when I conducted such exercises myself at the University of Strasbourg, I found them extremely beneficial. It has often bothered me that nothing similar is required of American students. In 1925, at the oral part of my examination, which was public and which took place at the Lycée Saint-Louis, a classmate of mine who had come to hear my lecture saw a member of the jury scribble a few words on a piece of paper, which he later left on the desk. My friend grabbed the paper and read it: "He will be a cabinet minister." Happily, this prediction, or rather this curse, did not come true.

Chapter III
First Journeys, First Writings

Like almost all the students whose families lived in Paris, as well as the few married students, I was not a boarder at the Ecole. Most of the students were, however, and they lived in Spartan conditions at which today's *Normaliens* would balk. They slept in large dormitory rooms, their beds separated from one another by thin screens which afforded the merest semblance of privacy: most horses are better off in their stables. The boarding students received room and board as well as a little pocket money, barely enough to buy themselves coffee from "the Baroness" – this was our nickname for the owner of the bar we used to frequent on the Rue Claude-Bernard, where more often than not we could be seen wearing our *"bonvoust"* attire – that is, the dungarees (the term "blue-jeans" was not yet in our vocabulary) that were issued to us for military training sessions. (The term *bonvoust* came from the name of a Captain who had conducted these sessions for a time.) Those students who wished to supplement their meager resources did so by tutoring *"tapirs"*, our slang word for *lycée* students seeking private tutoring; fortunately, they were in good supply. It has been said that, since those days, some *Normaliens* have made money in less conventional ways.

In contrast, the day students were entitled to a modest scholarship, theoretically equivalent to the cost of room and board for the live-in students. Living with my family, I had never had to resort to *tapirs*. My parents, though not rich, lived quite comfortably, as is normal for a doctor who is well liked by his patients; but it was a point of honor for me not to ask them for money.

In 1925, after the *Agrégation* examination, almost all of my class-mates left to do their military service as second lieutenants. Our class was the last one entitled to this rank without having to pass any examinations. I was still too young to do likewise. There were numerous scholarships in existence; they were given out parsimoniously – and I was not in very good graces with the administration of the Ecole. One of these scholarships, granted by the Sorbonne, was modest enough to excite little envy: this one was offered to me. It was intended for a student living in Paris, but I succeeded in arranging to spend an academic year in Rome, without worrying too much about whether the sum would be sufficient. I was beginning to feel rather hemmed in by my Parisian horizons.

I already owned Berenson's four volumes on Italian art, in which he catalogs the work of the great painters, and I used these for my visits to the Louvre. Over the summer, the Ecole allowed me to borrow Venturi's multi-volume history of Italian art, which in theory was not supposed to circulate outside the library. These readings furnished me with some artistic and linguistic background. Before leaving I spent two weeks with my family in Lanslevillard in the Haute Maurienne region of the French Alps, where I had my first experience of the mountains. There I took many long walks. At altitudes of 7000 feet or more, the air has a stimulating effect which I had never before experienced. Sometimes the whole family would go on longer excursions all the way to where the glaciers began. In those days, to go any further was supposed to require the services of a professional guide; this opportunity never presented itself.

My sister had this vacation in mind when she later wrote that contemplating a mountain landscape had once and for all impressed the notion of purity upon her soul. Of course, such a notion did not enter into my head: I was left with a totally different impression. Seeing the shafts of sunlight criss-cross in the distance at sundown gave me the idea of composing on several planes simultaneously. One should write, I thought, in such a way that the reader's mind is drawn behind the immediate subject, toward a plane behind it, and then on to other still more distant perspectives. There is nothing original about this idea; I might well have picked it up elsewhere – for example, in some of the Italian paintings I was preparing myself to see shortly. But it came to me there, in the mountains, whereas in my sister these same landscapes inspired radically different reflections.

In the course of my walks, I would often stop to open a notebook of calculations on Diophantine equations. The mystery of Fermat's equation attracted me, but I already knew enough about it to realize that the only hope of progress lay in a fresh vantage point. At the same time, reading Riemann and Klein had convinced me that the notion of birational invariance had to be brought to the foreground. My calculations showed me that Fermat's methods, as well as his successors', all rested on one virtually obvious remark, to wit: If $P(x,y)$ and $Q(x,y)$ are homogeneous polynomials algebraically prime to each other, with integer coefficients, and if x,y are integers prime to each other, then $P(x,y) Q(x,y)$ are "almost" prime to each other, that is to say, their G.C.D. admits a finite number of possible values; if, then, given $P(x,y) Q(x,y)=z^n$, where n is the sum of the degrees of P and Q, $P(x,y)$ and $Q(x,y)$ are "almost" exact n-th powers. I attempted to translate this remark into a birationally invariant language, and had no difficulty in doing so. Here already was the embryo of my future thesis.

In October it was time to leave for Italy. I took a leisurely route, through Milan, Bergamo, Verona, Vicenza, Padua, Venice, and Florence. The profession of tourist has its own methodology, and needs to be learned. It happened that I showed a natural aptitude for this profession, or better, for this art. This aptitude, combined with a certain linguistic facility, has contributed in no small way to my happiness in this life. I realized that in Italy, the thing to do was to make oneself Italian: herein lies the secret. Because my meager stipend forced me to husband my resources carefully, I quickly learned to scout out the neighborhood, the hotel, the restaurant, or the dairy shop where I could find, at the best possible price, what wealthier travelers often miss out on entirely. In Milan, my status as a foreign student enabled me to obtain the *tessera* which grants free admittance to all museums in Italy, thus freeing me from worry about an item that would otherwise have taxed my budget heavily.

Arriving in Rome after a month of traveling, already able to speak Italian passably, I felt quite at home there, all the more so because the renowned Vito Volterra extended me a fatherly welcome. Though probably a less universal mathematician than Hadamard, he was an admirable man in all respects. The king had named him Senator for life; he and Croce were the two senators who, until the bitter end, come what may, voted against the Fascists. In 1925, Giacomo Matteotti had just been assassinated. On the quay of the Tiber where he had died, flowers would be heaped up every night, only to be swept up by police patrols each morning. An air of unreality seemed to hang over political life. Perhaps one strong shove would have toppled the regime, but it never came. I remained an interested spectator of all of this, but I did not worry about it. After a few tries, I ended up boarding with a working-class family, not far from Santa Maria Maggiore. I had a fairly nice room there. We ate in the kitchen, where the mother, a generously proportioned woman of old Roman stock, prepared homemade *pasta, gnocchi,* and *carciofi alla romana,* as tastily prepared as in the better restaurants. There were no bathtubs, nor was there heating, but this was not a problem as I could take refuge in the university library whenever the cold drove me from my room. With the excellent Touring Club guide in hand, I visited Rome. The six months I had to spend there were by no means too long for that purpose. Rome is indeed the eternal city: all epochs are joined together there in a whole harmoniously fused by the patina of time. Dazzled though I was by the masterpieces of classical art, I fancied myself a Futurist, and frequented the hang-outs of the Italian Futurists, for which I reaped very little reward except the following: at the home of Filippo Marinetti, the picturesque leader of the movement, I saw works by Umberto Boccioni, including his admirable triptych "The Farewells," "Those who go," and "Those who stay." I do not know where the triptych is now; I only hope the three panels have not been separated. I learned that Boccioni had died very young during the war, in 1916.

With Edoardo Volterra, Vito's older son, himself destined for a brilliant career, I attended concerts in the *Augusteo,* with standing-room tickets at the highest circle. After the concert, we would escort each other home, back and forth for many hours; soon we were like brothers. With him I visited the famous Benedictine monastery in Subiaco. Perched on a mountain slope high above Rome, it was one of the cradles of printing in Italy. I hope that the monks are still there, and that their hospitality is still offered as freely. As for the *Augusteo*, the concert hall bizarrely housed in a cylindrical building inherited from Roman antiquity, it is no longer there – another victim of the Fascist regime. "That man," Vito Volterra would later say of Mussolini, not without exaggeration, "has done more damage in Rome than in Madrid."

Overwhelmed by so many new impressions, I worked in moderation, or rather, I dreamed about mathematics as I strolled through the city. I believed I had some ideas on linear functionals. However shapeless these ideas were, Volterra would lend his ear with tireless patience. I often saw his prize pupil, Fantappié, who remained his favorite until one day when he came to Volterra to sing the praises of the anti-Semitic legislation that Mussolini, following in Hitler's footsteps, had just introduced in Italy. Volterra was Jewish, nor could anyone be unaware of this fact. "How is it possible," he said, recounting this episode, "that I did not have the presence of mind to throw him down the stairs?" But in 1925, the situation had not reached that point. Fantappié was already wearing the Fascist *distintivo* ribbon in his buttonhole, but we were good friends. We would discuss functional calculus. Proposing from the start that we use the familiar form of address, he put me on the spot: my knowledge of Italian had not yet been extended to the second person singular.

In Rome at that time there was another *Normalien*, Guérard des Lauriers. I think Vessiot had recommended him for one of the prestigious fellowships from the Rockefeller Foundation. Both of us regularly visited the churches of Rome, but there was little chance of our running into each other: I was after medieval and Renaissance masterpieces, whereas he was seeking indulgences, which in the so-called "jubilee" year of 1925 were apparently available on particularly advantageous terms. I sometimes asked him, "Do you pray for me, Guérard?" – to which he would answer in all seriousness, "Yes, Weil, I pray for you." For a long time after Rome, I had no news of him. I later learned that he had become a Dominican. In a response he has written to my sister's *Lettre à un religieux*, he takes her to task, on the authority of several ecumenical councils, for suggesting that the eternal fate of a baby who has died unbaptized cannot differ much from that of one who has been baptized. Alas, the second Vatican Council found less favor in his eyes. I was told that he had fallen into integrism, and that he had died excommunicated.

Guérard and I were not the only foreign mathematics students in Rome: Mandelbrojt and Zariski were also there. Zariski had just been married the previous year. Severi was kind enough to offer us a few lectures in which he sketched out the basics of the theory of algebraic surfaces. He was Aretine, and eloquent; his Italian was more beautiful than any I have ever had occasion to hear. Listening to him, I had trouble paying attention to the substance of his talks, of which I remember little; at most, I benefited from them only unconsciously, if at all. The same applies in even greater measure to a series of informal talks on the same subject for which Enriques invited us to his home. Lefschetz made a brief trip to Rome, and I asked Severi what he thought of Lefschetz's work. *"E bravo,"* Severi told me, meaning more or less "he is talented": the word *bravo,* as far as I can tell, has no equivalent in other languages. "He is no Poincaré," Severi added; Poincaré was an eagle, *"un'aquila"* – and at this he raised his hand high; Lefschetz was a sparrow, *"un passero,"* and he lowered his hand halfway; but Lefschetz was talented, *"è bravo però, è bravo."*

It happened that a far less significant occasion, the appearance in Rome of an American lady-mathematician whose name I have forgotten, had more decisive consequences for me. After a lecture that she was graciously invited to give on the subject of Diophantine equations, she handed out offprints. In her bibliography she cited Mordell's now cele- brated paper of 1922. The question was too close to my own reflections of the previous year not to excite my curiosity. I read Mordell, therefore, but for the time being I did not turn it to account. Italy held too strong an attraction for me. This first Italian journey of mine ended felicitously with an Easter vacation in Naples, Ravello, and Sicily, followed by the return trip from Rome to Paris, in the course of which I lingered in Umbria and Tuscany. The harvest I reaped from these travels was not such that I could have reported on it at great length; still, if anyone had told me that I had wasted my time, I would have been greatly surprised.

By then the time had come to make plans for the following year. The Rockefeller Foundation, which was then only beginning its inter- national programs, had given priority to a program of travel grants for which mathematicians were long eligible. The foundation had adopted the following motto: "Raise the peaks; seek not to fill in the valleys." Its director for European programs, Mr. Trowbridge, was a very able man who scoured the world to find the most qualified candidates. In 1926 he consulted Volterra, who straightaway suggested nominating me for a grant. After Italy my wish was to visit Germany. Nothing could have been simpler. All I had to do was to select the place where I wanted to be and the person with whom I wished to work, and to obtain this person's consent. I chose Courant in Göttingen, because of linear functionals. Soon thereafter, I made known my wish to include a stay in Berlin during the year; this request presented no problems either.

On top of this, another project was added. Villat, a hydrodynamicist and the director of the *Mémorial des Sciences Mathématiques,* prided himself on taking an interest in young people. He knew me slightly. For quite some time he had been waiting for a monograph from Mittag-Leffler on polynomial series, to be published in the *Mémorial.* One or two years earlier, he had sent a *Normalien* to Stockholm for the purpose of writing it. Thinking he had reason to believe that the article was close to completion, he suggested to me that I finish it. There was an empty space of nearly two months in my program, between the winter and summer semesters at the German universities. I proposed to spend just one month and no more in Stockholm, and to do my best to finish it in that time. He agreed, and so did the Rockefeller Foundation. It seemed to me that my ideas, or what I thought of as my ideas, on linear functionals might well be applicable to polynomial series.

Once again taking a roundabout route – through Belgium, the Netherlands, and the Rhine Valley – I arrived in Göttingen in November 1926, in time for the beginning of the winter semester. Right away I went to see Courant, who was expecting me, in his large house on the Friedländerweg, quite near the main university building where most mathematical activity was concentrated at the time. This was before he had had the mathematics institute – over which he presided only briefly, because of Hitler – built *(sic vos non vobis...).* It has sometimes occurred to me that God, in His wisdom, one day came to repent for not having had Courant born in America, and He sent Hitler into the world expressly to rectify this error. After the war, when I said as much to Hellinger, he told me, "Weil, you have the meanest tongue I know."

Courant extended a cordial welcome to me. His first question was whether I played the cello: every year his wife organized a small chamber music group, and that year a cellist was needed. He was disappointed to hear that I didn't, but nonetheless he graciously invited me to be present when the group met. Then I started explaining my ideas on functional calculus. In short, I had in mind to extend Baire's classification to linear functionals; I had let myself be misled by the exaggerated importance the French school had accorded this classification, and I believed that some day this approach could shed new light on what were then referred to as "integral equations of the first kind." Courant listened to me patiently. Later on I learned that as of that day he concluded that I would be *"unproduktiv."* Leaving his house, I met his assistant, Hans Lewy, whose acquaintance I had made the day before. He asked me, "Has Courant given you a topic?" I was thunderstruck: neither in Paris nor in Rome had it occurred to me that one could "be given" a topic to work on. To those who asked him for a thesis topic, it was said, Sylvain Lévi would respond, "You have been taking courses with us for two years, and you haven't yet noticed that there are still questions to be answered?" I do not recall how I answered Hans Lewy; nevertheless, we became good friends.

I learned little from Courant and his group. Moreover, I hardly ever spoke to one of his students without the latter leaving me soon, saying "I have to go write a chapter for Courant's book." Hilbert was on the verge of retirement. He was the dignified chairman of the mathematical society's meetings, though he no longer came up with those caustic quips that people would repeat afterwards, imitating as best they could his Baltic accent. It is a pity these were not recorded before it was too late. The samples cited in English translation, in Constance Reid's biography of Hilbert, give only the palest idea of his biting wit. I remember nothing of the course he was still teaching that semester (for the last time, I believe), which I attended only in part. He was apparently trying to make the course of interest to physicists by discussing dynamics. As I found out much later, physics was thriving in Göttingen at that time: quantum mechanics was in the incubation stage. It is remarkable that I had not the slightest inkling of this development when I was there.

Emmy Noether good-naturedly played the role of mother hen and guardian angel, constantly clucking away in the midst of a group from which van der Waerden and Grell stood out. Her courses would have been more useful had they been less chaotic, but nevertheless it was in this setting, and in conversations with her entourage, that I was initiated into what was beginning to be called "modern algebra" and, more specifically, into the theory of ideals in polynomial rings. Paul Alexandrov could often be found at her house. When I told him that I had attended Lebesgue's course on "analysis situs" in Paris, he insisted on seeing my notebook from the course, and was sorely disappointed to find in it nothing that he had not already known for some time.

My mother's younger sister lived in Frankfurt. Her husband's Wagnerian first name, Siegfried, was paired with the rather non-Aryan family name Philippsohn. They had amassed a fine collection of old etchings. I spent my Christmas vacation at their house, not far from the Tiergarten, and took advantage of this trip to contact mathematicians in Frankfurt. Siegel's famous lecture, which concludes the third volume of his collected works, gives an idea of just how remarkable the atmosphere was there. Grouped around Max Dehn were Hellinger, Epstein, Szász, and Siegel, the latest arrival. It is not without feelings of gratitude and affection that I speak of this group here.

I have met two men in my life who make me think of Socrates: Max Dehn and Brice Parain. Both of them – like Socrates as we picture him from the accounts of his disciples – possessed a radiance which makes one naturally bow down before their memory: a quality, both intellectual and moral, that is perhaps best conveyed by the word "wisdom"; for holiness is another thing altogether. In comparison with the wise man, the saint is perhaps just a specialist – a specialist in holiness; whereas the wise man has no specialty. This is not to say, far from it, that Dehn was not a mathematician of great talent; he left behind a body of work of very high quality. But for such a man, truth is all one, and mathematics is but one of the mirrors in which it is reflected – perhaps more purely than it is elsewhere. Dehn's all-embracing mind held a profound knowledge of Greek philosophy and mathematics. Though less passionate, Hellinger was Dehn's kindred spirit. Certainly he could not have had the moral authority over his entourage that Max Dehn had acquired by his presence alone, but these two men seemed cut out to understand one another. They were competently seconded by Epstein and Szász, and all four of them took great pride in having Siegel at their side. Never again, anywhere, have I come across such a tightly-knit group of mathematicians.

A humanistic mathematician, who saw mathematics as one chapter – certainly not the least important – in the history of the human mind, Dehn could not fail to make an original contribution to the historical study of mathematics, or to involve his colleagues and students in this project. This contribution, or rather this creation, was the historical seminar of the Frankfurt mathematics institute. Nothing could have seemed simpler or less pretentious. A text would be chosen and read in the original, with an effort to follow closely not only the superficial lines of reasoning, but also the thrust of the underlying ideas. Here I am anticipating, for on my first trip to Frankfurt, the seminar had been suspended for the vacation period. It was only later that I attended it, on subsequent visits to Frankfurt, a place I made a point of visiting as often as I could. I am not sure whether it was already in the summer semester of 1926 that, during a seminar session devoted to Cavalieri, Dehn showed how this text had to be read from the viewpoint of the author, taking into account both what was commonly accepted in his lifetime and the new ideas that Cavalieri was trying to the best of his ability to implement. Everyone participated in the discussion, contributing what he could to the group effort.

During Christmas vacation in 1926, I met up only with Dehn and, for a few brief minutes, Siegel, who was returning to Frankfurt just as I was leaving. Siegel was already a legend; I had been told that he had drawers full of inspired manuscripts which he kept secret. In this regard Dehn explained the theory that was circulating in Frankfurt at the time:

mathematics was in danger of drowning in the endless streams of publications; but this flood had its source in a small number of original ideas, each
of which could be exploited only up to a certain point. If the originators of
such ideas stopped publishing them, the streams would run dry; then a fresh
start could be made. To this purpose, Dehn and his colleagues refrained
from publishing. But there was no moratorium on writing, still less on
presenting a friend with a manuscript for a special occasion such as New
Year's or a birthday.

Perhaps this theory had been invented above all to make sure Siegel
kept on producing work, at a time when the attitude reflected in the
American motto "Publish or perish" had invaded German universities and
increasingly filled Siegel and others like him with disgust. In any case,
Dehn had in his possession a manuscript on transcendental numbers which
Siegel had given him for his fiftieth birthday. I asked whether I could read
it. Dehn allowed me to do so, on condition that I read it at his home. I was
even permitted to take notes, with the further stipulation that I not transmit
them to anyone else. Apart from these conditions, Dehn did not insist on
secrecy regarding this work, which was later to become the first part of the
long paper published in the *Abhandlungen* of the Berlin Academy. But the
manuscript ended with the following carefully outlined words, which did
not appear in the printed version: *Ein Bourgeois, wer noch Algebra treibt!*
Es lebe die unbeschränkte Individualität der transzendenten Zahlen![1]

This episode was followed by another visit to Berlin, one that,
though fruitful for me, does not merit such a detailed description. I resumed
my reflections on Diophantine equations. Hopf, back from Amsterdam,
was teaching Brouwer's topology. He had helped arrange lodgings for me
quite close to where he lived, rather far from the center of town, and
together we would take the long tram ride to the university. One day I asked
him what he would do when he got tired of topology. He replied in all
seriousness: "But I'll never get tired of topology!"

Erhard Schmidt received me like the true aristocrat that he was. On
his mantelpiece was a marble bust of him. It resembled those of the ancient
Roman patricians, and he resembled his bust. His keen mind was nearly
comparable to Hadamard's – and to say this is no mean compliment. He
was quite willing to listen most attentively to what I wished to tell him
about functional calculus. He thought he saw some connection between
this and the as yet unpublished work of von Neumann, whom he suggested
that I should get to know. But von Neumann was not in Berlin, and I did
not meet him until some years later.

1   "It's a bourgeois, who still does algebra! Long live the unrestricted individuality
     of transcendental numbers!"

I also went to concerts. Thanks to Furtwängler I realized that Beethoven's symphonies are not merely a Sunday ritual, as they had seemed to me in Paris. I was eager to hear the famous Wilamowitz's lecture on Thucydides. As I was not a regularly enrolled student, I needed special permission from Wilamowitz himself. Though the Weimar Republic had abolished all honorific titles, I was advised to address him as *Exzellenz*. He received me at his home, and at first seemed overjoyed to have a French student visit him – the first since the war. He appeared somewhat disappointed to learn that I was not a hellenist, but he was no less well disposed to me for all that. He was nearing retirement, and I attended his very last class. That day, leaving aside the text he had explicated in the previous class, he announced that he was going to read and translate Pericles' Funeral Oration. As he did so he was visibly moved.

Thereupon I was to leave for Stockholm – not without some hesitation. One day in Berlin, after a lecture by L. E. J. Brouwer on intuitionism, we went to a café, as was the custom. I found myself next to Brouwer and confessed to him my doubts about the work that I had agreed to do for Mittag-Leffler, and which I didn't know how to get out of. "Nothing could be easier," he told me; "have a falling-out with him *(verkrachen Sie sich mit ihm)*." Indeed, for Brouwer, the solution was simple. I did not have a falling-out with Mittag-Leffler. My month with him was a very pleasant one for me, but of no avail for Villat and his *Mémorial*. I quickly decided that the draft prepared by my predecessor was useless.

I stayed in the Mittag-Leffler's own villa in Djursholm and had a number of conversations with him, all following the same pattern. They would begin in French, with a few remarks on power series and Weierstrass's enthusiasm for them. Then Mittag-Leffler would move on to his memories of Weierstrass, then of Sonia Kowalevska, and would spontaneously start speaking German; after which, growing tired, he would switch to Swedish. Finally he would give a start, saying, "But I forgot you don't know Swedish. We will continue our work another time." In fact, after two weeks, I knew enough Swedish to follow a conversation of this type. I began, in good faith, to write a *Mémorial;* but it was understood that I would spend no more than a month on it. At the end of the month, I had made so little progress that I dropped the project altogether.

One advantage of this trip was that I got to know Stockholm. It is a lovely city, and in the spring when the last of the ice melts it is irresistibly seductive. Moreover, Mittag-Leffler promised to print my future thesis in the *Acta;* and I had the privilege of spending not my days but my nights in his beautiful library where he kept, meticulously classified, a half-century's accumulated correspondence with all the great mathematicians of

Europe. It gave me a strange thrill to steal by night into the presence of Hermite, Poincaré, and Painlevé; within the intimate circle cast by the little lampshade, it was as if the outside world no longer existed.

I did not leave off my usual tourism: on my way back to Göttingen, I stopped off at Copenhagen, Lübeck, and Hamburg. Curious to experience a still rather uncommon mode of transportation, I took an airplane from Copenhagen to Lübeck. At that time the lesser commercial airlines were using the very small planes in which the passenger really experiences the physical sensation of flying – something that rarely happens nowadays. I began to combine this ordinary form of touring with a specifically mathematical variety. I had formed the ambition of becoming, like Hadamard, a "universal" mathematician: the way I expressed it was that I wished to know more than non-specialists and less than specialists about every mathematical topic. Naturally, I did not achieve either goal. At this time, congresses, colloquia, and other "symposia" were not yet in vogue; besides, it has always seemed more worthwhile to me to meet people, even scientists, in their natural habitat than in the midst of a randomly mixed crowd. Meeting someone in his own surroundings seems to make it easier to read his writings – or, sometimes, it becomes apparent that they are not worth reading. Despite all the errors to which this method exposes one, it actually saves considerable time.

In Hamburg, I was not able to find Artin. On the other hand, I did visit a large exhibit of Nolde's work, which left a lasting impression upon me. Later, back in Paris, I spoke enthusiastically of him to several friends who were devotees of modern painting. One of them, who ran a well-known art gallery, cut me short: "He has never been exhibited in Paris." Apparently, this settled his hash.

My return to Göttingen meant a resumption of Diophantine equations. Since my visit to Berlin, I had been able to formulate and to prove for algebraic curves what I called the "decomposition theorem" which, extended to varieties, was to make up the first chapter of my thesis. Suddenly, it became apparent to me that these same principles made it possible to see the true meaning of Mordell's calculations on elliptic curves, and that with this interpretation, I was in a position to extend these calculations to curves of genus greater than 1. I did not have Mordell's paper with me, and could not find it at the library. Ostrowski, then a *privatdozent* in Göttingen, was known not only for his talent but also for his erudition, which encompassed a wide variety of topics and made him a collector of offprints: I ran to his house, and indeed he did have Mordell's article, which he gladly let me borrow. In my exalted frame of mind, rereading it was a matter of minutes, and confirmed everything I sought in it. The next day, I took it back to

Ostrowski and told him that I could extend the results to include curves of arbitrary genus: here was the answer to a question Poincaré had asked twenty-five years earlier. I do not think Ostrowski believed me. I had in fact announced my results somewhat prematurely; my ideas needed a great deal of refining, which I was to do in the coming weeks, and it was to take me another year or two to develop them fully. Nevertheless I tried straightaway to interest Emmy Noether and her group of algebraists – to no avail, for number theory was not their forte, and their attention was monopolized by hypercomplex systems, that is, non-commutative algebras. Fortunately for me Siegel, with whom I was able to speak at length on a visit to Frankfurt, and to whom I explained my decomposition theorem, reassured me as to the value of this first discovery.

Hellinger, no doubt impressed by my youthful enthusiasm, also felt obligated to encourage me in my first stabs at the field of functional calculus. Together with his friend and collaborator Toeplitz, who was visiting at the time, Hellinger was busy writing the extensive section on integral equations for the German encyclopedia. This piece remains, at least from a historical point of view, a gold mine of precious information on the topic. When he was in Frankfurt, Toeplitz was a regular participant in the historical seminar, and the Frankfurt mathematicians looked upon him as one of their own. I like to imagine that they ended up thinking of me in the same way.

Regarding my return trip, I will mention only two things: a lengthy spell spent contemplating the altar by Tilman Riemenschneider in Rothenburg-ob-der-Tauber (or more precisely, in a small village chapel in the vicinity of this city), and my meeting in Munich with Hartogs, whose work, as I have said, I had admired since entering the Ecole. I was quite surprised to find a timid, self-effacing man with the look of a harmless rodent. I did not get much out of our meeting, but this did not diminish my admiration for him.

It was time to refine my ideas on Diophantine equations and to turn them into a doctoral thesis, a task to which I devoted myself in the summer of 1927 and the following year. Living with my family, I found a modest grant sufficient for my modest expenses. That summer I taught myself how to type. From the twofold perspective of saving time and of the increased independence conferred by mastery of this simple skill, the few weeks I spent acquiring it were perhaps the most fruitful of my life, and I have never stopped encouraging all young mathematicians, and indeed everyone else, to do the same. The era of computers was yet to come – and anyway I have always got along quite well without them. At the end of the summer, I was happy to return to Hadamard's seminar, Meillet's course, and concerts in Paris.

Writing a thesis, and even typing it myself, were not sufficient in themselves: I still had to have it accepted by the University. First I went to Hadamard for advice. I told him that my thesis solved a problem posed by Poincaré. I made the mistake of adding that I had hoped at the same time to prove what is now referred to as Mordell's conjecture, on the finite number of solutions to all equations of genus greater than 1, and that I had not yet succeeded. "Weil," he said to me, "several of us think highly of you; you owe it to yourself, when presenting your thesis, not to stop halfway through. What you say shows that your work is not yet mature." And so I renewed my efforts – but soon I decided to present my thesis such as it was. My decision was a wise one: it was to take more than half a century to prove Mordell's conjecture.

Number theory was a subject that was utterly neglected in France, but my thesis also discussed algebraic functions and Abelian functions. I thought it would be appropriate to go to Emile Picard. He knew me through having been my examiner for part of the *Licence* examination. He had been an excellent mathematician, but he was also an important personage, permanent secretary of the Academy of Sciences and a member of the Académie Française. Even before the war of 1914, the list of his titles and honors took up more than a page in the *Acta Mathematica*. I did not expect him to read my work; I hoped that he simply do what the circumstances required, that is, submit a favorable report. I had not realized his sense of professional responsibility. Reading a thesis thoroughly before approving it was a duty he took very seriously. Perhaps he was also annoyed by the audacity I showed, completely unawares, by calling him up one day on the telephone. Be that as it may, he telephoned me to say that he would gladly chair my thesis committee, but that I had to find someone else to submit the report.

In the meantime, Siegel had judged my work favorably. In Paris, everyone was telling me: "It's very interesting, why don't you go see so-and-so?" I had attended Lebesgue's courses. My topic was wholly unfamiliar to him, and it was not the custom then for a professor at the Collège de France to sit on a thesis committee; but he prided himself on departing from ordinary practice. I resolved to tell him: "My thesis is good, it has Siegel's approval; I can't go on like this. I'm not asking you to read this thesis; just agree, for the sake of form, to submit the report." This is more or less what I said to him at the Monday meeting of the Académie, where he had summoned me to speak with him. He told me to wait for him, went off to exchange a few words with Picard, and returned to me: "Do you have your manuscript?" "Yes." "Take a taxi to Monsieur Garnier's and tell him that Monsieur Picard wishes him to take charge of your thesis." Garnier had just been appointed to a post at the

Sorbonne; without Picard's decisive influence, it was said, another candidate (rumor had it that this was Fatou) could and would have received the appointment. I went straight to Garnier and delivered my spiel. He started to say, "That's interesting..." when I said the magic words. "Ah," he said glumly, "if Monsieur Picard wishes it..." Thus he was entrusted with the task of filing the report on my thesis, a task which he performed conscientiously and benevolently: he did not notice a few gaps in my proofs, but he did give me some useful advice on commas. This episode branded him as the official reporter for all theses on algebra and arithmetic. Chevalley's dissertation must have given him even more trouble than mine did.

Now all I had to do was get my thesis published. Mittag-Leffler had promised to give it space in the *Acta*. He had died shortly after making this promise, but his successor, Nörlund, agreed to honor it. The time had also come for me to do my military service. I had already been summoned before a draft board at the French consulate in Rome, where I had appeared stark naked along with a herd of French seminarians who were making fun of their German theology professor, trading ecclesiastical jokes at his expense. Because of my youth and my student status, my service had been deferred, as it was once again in Paris the following year; but this respite could not last forever.

My case was a troublesome one. I was the last survivor of a system wherein all alumni of the Ecole Normale automatically started at the rank of second lieutenant, after preliminary training periods which I had not gone through. I was advised to seek out Painlevé, who was President of the Chamber of Deputies; even more importantly, he had also been Minister of War. He did indeed give me a letter, addressed to a civil servant who bore with great dignity the impressive title of Director of the Infantry. It was decided that I would join the infantry, contrary to custom (scientific *Normaliens* usually served in the artillery), and that I would serve one month as a private, one month as a sergeant, and the rest of the year as a second lieutenant, one after the other without interruption, and all with a regiment of the Paris garrison.

This plan made for a very peaceful year, during which I lived at home with my parents. The 31st infantry regiment was not unfamiliar with *Normaliens,* and knew enough not to expect too much of them. I was not terribly impressed with the infantry officers. Some of them had risen from the ranks; all had served for the duration of the war of 1914, and most, I believe, had conducted themselves very honorably. But I was struck to see that these men, intrepid when facing real danger, seemed terror-stricken at the mere approach of some general or other conducting an inspection. Women consitituted, if not their principal activity, at least the principal subject of conversation among these men. At the officers' mess one day, a battalion chief poured out this melancholy confession: "I've had all sorts of women, even a Bedouin in Syria" (the rest of the conversation revealed that he had raped her when she had sought a remedy for toothache); "but I have never had a nun... and now it's too late." Another time, in the course of a march that was to include a number of days billeting in the Champagne region, I heard a sergeant-major lecturing his men (the word "man" meant "soldier": an officer is no longer a "man") as follows: "When you are billeting, behave yourselves around the girls. Sleep with them when you can, but show respect for them, always show respect!"

Such were the joys of battalion life, which I was in a hurry to put behind me. By finagling with some contradictory regulations, I managed to cut my service short by nearly two months, not counting a leave obtained on the pretext of having to correct the proof-sheets of my thesis. It was about time to find myself "a real job." A position was soon to be vacated at the University of Strasbourg, but it was to be filled by my friend Henri Cartan. For a year or two already, I had been saying to Sylvain Lévi that I would gladly go to India. One day in 1929 he telephoned me: "Are you serious about going to India?" "Of course." "Would you be willing to teach French civilization?" "French or any other, I don't care; to go to India I'd teach anything they want." "Well, get a cab and come right over to my place."

At Sylvain Lévi's, I found a tall man who filled the room with his broad frame, stentorian voice, and ringing laughter: his was one of those presences which makes itself felt anywhere. Sylvain Lévi and he appeared to treat each other with mutual respect. He was named Syed Ross Masood. It was to him, I later learned, that E.M. Forster had dedicated his *Passage to India*.

This man was the Minister of Education for the state of Hyderabad. While traveling in Europe, he had received a telegram offering him the presidency (with the traditional title of Vice-Chancellor) of the Aligarh Muslim University not far from Delhi. There could be no question of his turning down the offer; this institution, which enjoyed great prestige throughout Moslem India, had been founded by his grandfather, and was in an advanced state of decline. Masood was being granted full powers to rescue it from this condition, a task he considered to be a family obligation, as well, no doubt, as a stepping-stone to higher destiny – which he never did attain. When I met him, he was preparing to curtail his vacation so that he could return to Aligarh without delay. It was his wish that French culture find a place alongside English culture in India, and for starters he planned to create a chair of French civilization in his university. He asked whether

I would accept the position, with a salary of 1000 rupees a month. Having no notion of what the rupee was worth, I turned to Sylvain Lévi, who told me it was sufficient. In fact, it turned out to be more than enough, but this factor was not foremost in my mind. I consented to the terms. Masood told me that official notification of my appointment, as well as a travel allowance, would be cabled to me shortly.

Then, for several months, I heard nothing. From time to time I asked Sylvain Lévi if the whole thing was serious; without going out on a limb, he advised me to wait. Warned that there would be no French books in Aligarh, I made the rounds among publishers, solliciting donations primarily of literature and history books. I even obtained a small subsidy from the Ministry of Education. Finally I received a cable: "Impossible create chair French civilization. Mathematics chair open. Cable reply." I cabled. Shortly thereafter, I indeed received my travel allowance. Not for a second during my meeting with Masood had the subject of mathematics come up. Had Sylvain Lévi told him that I was a mathematician? It certainly seems plausible. But I was never told.

Chapter IV
India

I have never kept a diary, but while I was in India I sent a letter to my family every week, like clockwork. On a certain day of the week – I think it was Friday – the P & O (or, to give it its full name, the Peninsular and Oriental Navigation Company) steamer left Bombay carrying the "home-mail" from the English colonists in India to the mother country. Just like the administrators of the Indian Civil Service, or the Viceroy's splendid guard, this institution partook of all the majesty and the quasi-sacred character of the Empire, referred to in Anglo-Indian jargon as the "British Raj." In every post office, even in the most far-flung corners of India, barely touched by civilization, one could inquire about the home-mail, and straightaway one would be told the day and the precise hour of its departure. Leaving punctually on the appointed day, the mailboat sailed for Marseilles, and the mail arrived just as punctually at its destination, more regularly and perhaps faster than it does today.

Unfortunately, almost all of my weekly letters to my family seem to have been lost forever. Without these letters to jog my mind, I must be content to glean what I can from what memories I have.

To reach India, the traveler had a choice between the P & O and the Italian boats of the Lloyd Triestino line. From the Indian students I knew in Paris I had learned that the P & O was patronized above all by English bureaucrats, and that the atmosphere was rather stuffy and (to use a word not yet in fashion at that time) "colonialist." To become better acquainted with the English than I was then, I thought it wiser to await such a time as I might visit them on their own territory. It was hardly to rub elbows with the least interesting specimens of the English that I was going all the way to India. I therefore chose an Italian ship, which I boarded at Genoa at the start of 1930.

At that time, these ships were still rather low in tonnage; from Genoa to Bombay, the voyage took just two weeks. Unless one made a special effort to isolate oneself, fifteen days allowed plenty of time for everyone on board – passengers, officers, and sailors – to become acquainted with one another. There were no movies, but there was "deck-tennis" played with rope rings fashioned by the crew. The atmosphere was merry and relaxed. Naturally a number of shipboard romances developed, only to be dissolved as soon as the ship approached the docks in Bombay; but outwardly, utter decency prevailed throughout. These boats still made use of the *table d'hôte* system, long since outmoded in European hotels:

A.W. in Aligarh with Vijayaraghavan and two students (1931)

the captain sat at the head of the table in the first-class dining room, and other officers presided over the second-class tables. We were served excellent Italian food. Outside mealtimes, one moved freely from one class to another without eliciting objections. As we passed through the Suez Canal and the Red Sea, we all had our mattresses brought up on deck. One night, trying to get a breath of fresh air, I hoisted myself up onto the weather-cloth over the promenade deck, where a sudden gust of wind nearly put an early end to my career as I slept.

    The passengers came from all over the world; there were even two natives of Chicago, off to India to sell refrigerators. They gave me my first taste of American civilization. "Prohibition is a good thing," they said, "we are all for it. But good health requires a drink once a week." For them "once a week" meant a continuous bout of whisky drinking from Saturday through Monday morning. There were also steerage passengers: these were mostly young Germans who, relegated by the developing crisis to the ranks of the unemployed, had converted to Buddhism. Dressed in yellow, their heads shaven, they were bound for a monastery in Ceylon, their expenses footed by the Buddhist community. They spent their time quietly boning up on their new religion by reading Buddha's sermons in German.

I did not linger in Bombay. Despite the view of the Indian Ocean, Malabar Hill, and Elephanta Island, it is an unattractive city. I was initiated to Indian cuisine at the house of one of the great Muslim families of the city, whose son I had met when he was studying in Paris. As everyone knows, Indian food is very spicy, primarily from the use of chili peppers, which the Portuguese imported from America to India in the sixteenth century. As it happened, I was invited to a wedding feast. In this family, which was quite Westernized, the custom of purdah (literally, "curtain" or "veil"), whereby women were kept in isolation, was not observed. I had the honor of being seated next to the hostess, who kindly reasssured me that for my sake she had seen to it that the dishes were not too highly spiced. Nevertheless the food was hot enough to smart, and I had to make an effort to conceal my discomfort.

These friends also helped outfit me for my stay. They could not repress a hint of irony on seeing the cork helmet with which I had thought it necessary to equip myself in Paris. But they made me purchase the indispensable bedding which was to become my travel companion for as long as I was in India. It was made up of a thin mattress, a pillow, a pair of sheets, and a mosquito net, all rolled up in a heavy-canvas carrying case. On a train, one would spread it out on the berth; when visiting friends, one would set it up on the cot provided by the host. The trip from Bombay to Delhi presented me soon enough with an opportunity to practice using this bedding, as well as the chance to become acquainted with the pervasive dust from the Indian plains and the clamor of crowds in the stations. It is not far from Delhi to Aligarh, where I was met at the station. Sweet-smelling garlands were placed around my neck. Though surprised and somewhat embarrassed, I was careful not to let it show. "What a pity that you didn't come a day earlier," I was told, "our annual flower show has just ended." "No matter," I replied, "I'll see it next year." My spontaneous reply pleased my interlocutor, who told me: "You are already Indian."

Thus, in January 1930, began a stay that was to last more than two years, and which left me with a prodigious treasure trove of impressions. This plethora of stimuli can be compared only to what a small child takes in during the first years of life on this earth – except that I was fully cognizant from the start, and I still have vivid memories. In Sanskrit, the brahmin is said to be "twice born," *dvija:* the second birth is conferred upon him by the brahmanic cord. It was not simply as a joke that, shortly before I was to return to France, my friend Vijayaraghavan girded me with such a cord. Was it not the symbol of my second birth?

In Aligarh I found Masood just as I had seen him in Paris. For several weeks, I stayed in his house and shared his meals, and thus had a chance to become accustomed to Indian food, for which I soon developed a taste. I met a young zoologist, Baber Mirza, fresh from the University of Frankfurt. He claimed to be of Mughal origin. Masood had met him in Germany and recruited him for Aligarh on the spot. Baber Mirza and I became good friends and were to share a house at the edge of the campus. He was not married, and complained sorely about his bachelorhood. The next year, he took advantage of his vacation to return to Germany, bringing back with him a prettyish, seemingly guileless little German girl who, as I was to learn much later, became a Nazi sympathizer and persuaded her husband to support the pro-Japanese party.

While waiting for our house to be readied for us, I had to familiarize myself with my new surroundings as well as my new duties. Inexperienced as I was, and having had Sylvain Lévi's apparent respect for Masood impressed upon me, I expected the latter to guide me through the many obstacles that lay before me. In the Muslim community in India, the Vice-Chancellor of the University of Aligarh was an important figure, and Masood played the part to the hilt. Seeing him preside over a group of distinguished dinner guests and regale them with an endless tide of anecdotes in English and Urdu was a spectacle of which one could never get enough. It was only much later, after returning to France, that I understood just how much this brilliant façade served to cover up a gaping void.

As it happened, Masood had been appointed to his position with the principal mission of making a clean sweep of the faculty. Despite the presence on the faculty of a few intelligent men (among these were a historian and a philosopher), the teaching staff was woefully unqualified, especially in scientific disciplines. Masood had been granted full powers to transfer or fire faculty members as he saw fit, with or without prior notice. This is exactly what had happened with the position I now enjoyed: my predecessor, a bearded and bigamous Muslim (bigamy was legal for Muslims) had been shunted off into the insignificant position of Director of the College of Education. He had studied mathematics in Germany, where he had somehow managed to get himself a doctorate, by hook or by crook. In Germany as in France, it was then commonly thought that granting diplomas to Indians or other "natives" was not a matter of great consequence, since they would soon be leaving Europe, and once home they would act as walking propagandists for the country where they had studied.

Everyone felt threatened in this setting, which was rife with intrigue. Naïve though I was, it did not take long for me to catch on. One of my colleagues was a pleasant man of Persian descent who taught Persian language and literature. More than anyone else he bore the brunt of people's suspicions. When I asked why, I was told, "What do you expect? He's a Persian."

As I was to realize later, the university (no different from other Indian universities in this respect) bore a strong resemblance to the princely courts that the English had taken pains to preserve in India, and which disappeared only after India's independence. The following year I had occasion to visit a typical specimen of such a court, in Rampur, where I was drawn by the reputation of the *nawab's* musicians. Everything there hung upon the whim of the prince (or the "ruler," to use the empty title he preferred), and plots were constantly being spun and unraveled around him. The same was true in Aligarh, with countless intrigues revolving around the person of Masood. For some time, I found myself among the court favorites and was treated accordingly. Needless to say, this situation could not last forever.

Everyone knows that no university institution is immune to intrigues of all sorts; but during my stay at Aligarh Muslim University I saw more, and more sophisticated, intrigues than during the entire forty years of the rest of my career. I have only indistinct memories of them, but I do recall being told a masterpiece of the genre, concerning not a university but a court. This episode, it seems to me, rivals the story of the renegade of Avignon, which Rameau's nephew recounted with such gusto.[1] It is hardly a digression from the matter at hand.

In the state of Hyderabad the prince (the *Nizam*) had a favorite wife or concubine: naturally she had enemies, and they formed a party against her. The woman had a weakness: she had not borne a child. To have a child by the Nizam would have secured her position once and for all.

India is richly endowed with holy men whose talents extend to blessing the unions of barren women. Though the results may sometimes be brought about by natural means, a more delicate approach was called for where a consort of the Nizam was involved. A holy man was lodged in the outskirts of the city. By means of rumors cleverly planted among the townsfolk, his reputation was built up, with plenty of time being allowed for the effect to take hold. It was calculated that, sooner or later, the woman would send a messenger to the holy man, to obtain the secret for the much-desired pregnancy.

And so she did. The holy man's instructions were easy to follow. The woman was to have the Nizam dine with her, and for dessert to offer him delicacies prepared with her own hands using a magic flour provided by the holy man. Afterwards, she had only to keep the Nizam with her for

---

1   *Translator's note:* the story is told in Diderot's *Le neveu de Rameau.*

the night, and pregnancy would ensue. The dinner took place as planned. At the critical moment, servants rushed in crying, "His Highness is being poisoned!" The cakes were offered to one of the Nizam's favorite dogs. The animal dropped dead before their eyes. Naturally the holy man vanished without a trace. As for the lady, I will leave her fate to the reader's imagination.

Naturally I witnessed nothing this dramatic at the university; but I will add a much more innocent anecdote, also from outside the university, but which took place in the town of Aligarh. One of my colleagues, an elderly Muslim who was teaching chemistry, was extremely fond of guavas. Each year, in guava season, he would order several cases of high-quality fruit to be shipped to him from quite some distance away. One year, he did not receive his shipment. After making inquiries, he came to suspect the employees at the Aligarh railway station. The following year, he went to the source of the fruit and injected a powerful purgative into each guava that was to be sent to him. Then he returned home and, at the appropriate time, appeared at the station. He found the entire station staff doubled over with cramps.

Be that as it may, what the university expected me to do in the short run, apart from teaching rather low-level mathematics courses which demanded little effort, was to produce a report on the teaching staff in my department. The continuation or termination of each individual's contract hinged upon this report. As in every department, the mathematics faculty, designated by terms imported from England's traditional education system, consisted of one professor and one reader, with two lecturers at the bottom of the ladder. The reader was a Hindu from Bengal, a diminutive and obsequious man, who had studied in Calcutta with Ganesh Prasad. This professor, entirely forgotten today, had placed the products of his instruction all over northern India. Of the two lecturers, both Muslim, one enjoyed great popularity among the students because of his willingness to help them – this along with a few scattered successes on competitive government examinations – and his beard, which was dyed red like the Prophet's. The other claimed to have undertaken the study of a mathematical work in Arabic (the *Qanun Al Masudi*), of which a celebrated manuscript had formerly belonged to Aligarh University but had been sold off to Germany; reportedly they had kept a photocopy, which I never saw. The reader had one advantage over the others: in Calcutta he had learned the notion of uniform convergence. With this exception, all three were equally devoid of merit.

At the age of twenty-three, therefore, I held the fate of these three pathetic characters in my hands, and their behavior toward me reflected their awareness of the situation. Furthermore, I was given only a few weeks to come up with my report. No doubt the problem was as much a human as a scientific one, but Masood, pleading his utter ignorance of mathematics, refused to give me any directive. Needless to say, I did not even consider consulting my predecessor. Besides, the programs of instruction and examinations seemed to me to be in urgent need of overhauling, but my modest stabs in that direction were hardly welcomed, and were to a large degree at the root of my ultimate downfall: they threatened to rock the boat too much.

I would gladly have rid myself of all my three acolytes in one fell swoop, but no one at the university was able to tell me whether I would be able to find replacements. The only Indian mathematician whose name I knew was Ramanujan, and he was long since dead. Finally, I resigned myself to keeping the reader. This was my first mistake. Of the two lecturers, I fired the second and granted the first, the red-bearded man, a year's reprieve. He appeared to accept his disgrace philosophically, but it would have been more politic to retain him in his functions which the poor man fulfilled as best he could.

During this time, the house Baber Mirza and I had taken was being readied for our occupancy, with furniture I had had made following Parisian designs. Fortunately, it was not a recently-built structure. Thick mud walls, a substantial thatch roof which was home to a family of mongooses, a large terrace which extended all around the roof, several verandahs, spacious high-ceilinged rooms: everything was designed with a view to diminishing the effects of heat in the summer months and conserving it during the winter frosts. There was nothing in the way of what we (rightly or wrongly) refer to as "conveniences": no bathtubs, no running water. Asians traditionally look askance upon what they regard as the European practice of simmering in one's own filth. Here daily bathing meant dousing oneself with pitchers of water brought by the water carrier (the *bhishti*) and heated in the kitchen. Cooking was done on wood fires at the far end of the garden (or the "compound," in the Anglo-Indian parlance of the time). This was where the servants lived with their families. I never found out how many of them there were, nor did I have much to do with them; I left these matters to my friend Baber. They were living in rather tight quarters, perhaps, but I do not believe that they were unhappy; the very fact of working in the employ of a European, or of a Westernized Indian, was an enviable privilege which meant better wages and an elevated status in the complex hierarchy of their social group. Also at the far end of the garden, it was said, were a couple of cobras, not to mention at least one coral snake, which Baber found inside his slipper one day. A number of frightful anecdotes involving such snakes made the rounds, at least among the English. An English judge was said to keep a supply of serum at his home, ready to be administered in case of need. Called out of his house one day, he found a *tonga*, a sort of light two-wheeled vehicle in which the passenger rode with his back to the (usually foul-smelling) horse and to the driver. On the seat a man lay motionless, clearly in a bad way. The man who had driven him said that he had been bitten by a snake. "He is too far gone for me to help," said the judge; "take him to the hospital immediately." "*Sahib*, we were just there; they turned us away." "Why did they do that?" asked the judge. "*Sahib*, they told me he was dead." As for my cobras, I never saw them; I only saw the skin of one of them, caught in a bush. Because of these snakes one had to be somewhat cautious walking on the garden paths, especially in the rainy season. Apart from this, they were no cause for concern. They were moreover sacred animals, which our Hindu servants would have taken pains not to harm.

There was no electricity in our house when we moved in, and therefore no electric fans. Instead there was the *panka*. This was a piece of cloth suspended from the ceiling and set in motion by means of a long string in the lazy hands of an unfortunate boy stationed in the corner of the verandah. As the hot season approached, it became necessary to take a siesta, and for some time I relied on the *panka* to supply relief. Naturally the boy would doze off. No longer feeling the tepid air wafting over me I would wake up and cry, "*Pankevale!*" He would stir and, for a time anyway, set the *panka* in motion again. Although I am not a great fan of progress, I was pleased when electricity was installed in our house.

In front of the house Baber planted roses, which thrive in the climate of northern India. Later on my friend Zakir Husain, when he had become the president of India, showed me a marvelous rose garden that he had created behind the presidential palace. We also had some trees inhabited by parrokeets and those little squirrels that are such a common sight in the United States, but which were new to me at the time. The nights passed in a profound silence, interrupted only by the mournful cry of the *chowkidar* or night watchman. This post, I was told, could be held only by the members of a caste of professional bandits who had "reformed," preferring a regular modest income to the hazards of their former profession. Each one was responsible for a group of houses, from which he was supposed to fend off nighttime intruders with his cries. The louder and more frequent the cry, the better the crier thought he was earning his salary. I politely requested that mine moderate his zeal. Afterwards my nights were indeed more tranquil. As the hot season came on, I had my bed carried onto the roof. Never before had I fully experienced the beauty of sleeping under the stars. As I was later to observe in Brazil as well, the night sky in tropical regions is sublimely clear and pure, and the stars there look brighter and more numerous than in our temperate climates. At least it used to be so; nowadays, people say, pollution has changed all that.

It is perhaps trite to quote Kant: "the starry sky high above our heads, moral law deep in our hearts." As for moral law, the phrase, alas, sounds hollow, whatever the respect due the old master of Königsberg. But I have never been able to see the starry sky, *cum tacet nox*,[1] without being moved by the sight. Already in Italy it had appeared more luminous to me than in France. How then can I find words for the effect the Indian sky had upon me when I had the chance to contemplate the spectacle? And how can I describe the moonlight in these latitudes, without verging on the ridiculous? It is so strong one can read a newspaper by it – though that would be a sorry use for it. Travelers and guides rhapsodize about the sight of the Taj Mahal by moonlight. No doubt it merits the epithet "magical," which has of course been applied to it; but I have never really had a taste for this bastardized offspring of Italian baroque grafted onto the ostentatious whims of a Mughal despot. One full-moon night in Aligarh showed me something even better: this was Fatehpur Sikri.

The story is well known: Akbar wished to establish his capital there; for lack of water, the site was abandoned, but not before Akbar and some of his courtiers had built, for themselves and their wives, palaces which this very lack of water has helped preserve. Fatehpur Sikri is not far from Agra, and Aligarh is halfway between Agra and Delhi. These distances are not great by car.

When I was in Aligarh, there were no taxis or public transportation; instead, there were *tongas*. When a couple of friends and I decided to spend a full-moon night at Fatehpur Sikri, the *tongas* would have been of little use to us. We appealed to the kindness of a rich Muslim landowner who congratulated himself on his good relations with the university. He had an automobile and a chauffeur, which he graciously lent us.

Nowadays Fatehpur Sikri, rightfully placed under the protection of the archeological authorities, is surrounded by walls. The *Fateh Darwaza,* the splendid victory gate, is equipped with a ticket window; admission is by ticket only and is restricted to certain hours. When I visited the site, it was open at all times, to any and all visitors. Few actually came. A modest inn with a few beds made it possible for us to spend the night. Until the early morning hours we wandered around the ghost town. We could not get enough of exploring the galleries and the women's quarters, enclosed by elaborately carved stone fences which once allowed them to gratify their curiosity about passersby, without themselves being seen. The moon was

---

1   "When the night is silent": Catullus' beautiful lines are well known: *Aut quam sidera multa, cum tacet nox, / Furtivos hominum vident amores* (Cat. VII, 7-8).

reflected on the tile roofs and pierced through the windows' latticework, illuminating the walls with its surreal light. I do not remember how late it was when we finally tore ourselves from this spectacle, which today no traveler can witness. Or has some enterprising Indian promoter already organized a "Sound and Light" show there, at so much a head?

Aligarh itself did not boast such attractions or, for that matter, attractions of any sort. I went into town only to take the train. But I soon fell in love with the broad northern plain and its vast horizons, broken only by irrigation canals and groves of banyan trees with multiple trunks where, as dusk fell, countless little green parrokeets would come to perch and chatter. Often I would go for long walks, sometimes with one or another of my students. From time to time we would skirt a village, with its humble mud-walled shanties, covered with cowpats or buffalo dung set out to dry for use as fuel.

But the hot season was approaching, and with it vacation time, scheduled that year for the months of April, May, and June. I had planned to spend them in Kashmir with an Indian colleague, a friendly young scholar of Arabic who was giving me Urdu lessons for the fun of it. On the other hand, Masood's clean sweep had taken effect, leaving a large number of positions open. One of these was my lectureship. According to procedure, each position was advertised in the newspapers. A preliminary selection of applicants was made on the basis of credentials, and the selected candidates were summoned before a committee formed for the purpose of making the final choice. These committees were to meet at the end of May. Since my department was involved in a search, I was neither willing nor able to absent myself from this process. It was agreed that I would interrupt my vacation – however unpleasant this prospect was in the worst of the hot season – for a quick trip to Aligarh. As for the preliminary selection, Masood's assistant, a Scotsman with the title of Pro-Vice-Chancellor (who was to die of sunstroke before the summer's end) assured me that he could easily take care of it. I did not trust his judgment, and insisted that a complete list of the candidates with an abbreviated list of their qualifications be sent to me in due course in Kashmir. There were over a hundred applicants. The most absurd application, from the French trading post in Chandernagor, came accompanied by a personal letter in French which ended thus: "...I believe that my personal status as a French citizen must prevail, when the choice is in the hands of a French gentleman" (for "gentleman" the misguided soul had used *"gentilhomme"*). The list reached me in Gulmarg ("valley of roses"), a mid-altitude mountain resort from which a splendid view of Nanga Parbat ("naked mountain"), one of the most beautiful peaks of the Himalayas, could be enjoyed. It took me no time at all to see that this list contained the name of only one mathematician, in the sense I ascribe to the word: this was a pupil of Hardy's, by the name of Vijayaraghavan, who had to his credit several articles on approximation and Tauberian theorems, but no degree, and who was therefore not on the Scotsman's short list. I ran to the nearest telegraph office and told him to include Vijayaraghavan in the list. As soon as I saw him in Aligarh, I was positive that my choice was the right one. My only regret was not to have gotten rid of my Bengali reader, for Vijayaraghavan would have been perfectly qualified for that position. His impeccable Oxford English, which he spoke with a slight Madras lilt, and his no less impeccable turban of raw silk made him acceptable to everyone else as well. I was able to take the train back to Rawalpindi, where I had to make the connection for Kashmir.

What can I add to the concert of poets and travelers who have sung the praises of Kashmir? I had great expectations, and they were more than fulfilled. My friend Abid and I began by renting a houseboat, one of the barges outfitted for tourists and manned by a suitable crew, which at a modest price provided the most enjoyable means of exploring Srinagar and the chain of lakes nearby. Kashmir is on the ancient caravan route that, from time immemorial, has connected Central Asia to India on the one hand and to Arabia on the other. It was once inhabited by sophisticated Mughal princes whose love of lush landscapes led them to embellish these already incomparable sites with their own parks.

Kashmir was then a "native state", part of the British Empire, of course; the primarily Muslim population was supposedly governed by a Hindu maharajah. The country was extremely peaceful, and tourism was its major source of income. Tourists were exploited mercilessly but, as in Italy, always with a smile, and any tourist who could defend himself was treated none the worse for it – quite the opposite, in fact. It was in Kashmir that I earned my stripes in bargaining, that indispensable part of Oriental life, my initiation into which had begun in Italy: as the Italians say, the East begins in Naples. I spent several days with a dealer in antiques and carpets, whose shop (which was also his home) was chock full of marvels. As was customary, I chatted with him as he offered me tea and unrolled one sumptuous carpet after another for his visitor and prospective client. The art of bargaining requires first of all that the customer feign indifference with respect to what he most wishes to acquire. In India, the story is told of a village story-teller who decided one day to switch professions and open a boutique at the bazaar. His unhappy admirers decided to force him back to his true vocation. They agreed among themselves to buy from him without any bargaining. After a few weeks, he succumbed to boredom and closed up shop. The same was hardly to be feared from my rug merchant. When I finally left him I had gained possession, at what seemed to me to be moderate prices, of two or three splendid prayer rugs from Persia or

Central Asia. They were later to suffer a sorry end, devoured by moths in storage during the war.

After the lakes of Srinigar, we went to Gulmarg with its unforgettable views of the Himalayas. Then, as the end of May came with its overpowering heat, I made my brief trip back to Aligarh. For the second part of our journey, we had planned to trace the upper valley of the Indus to the Zoji La pass and even a little beyond – as far as we could get without obtaining a special permit from the authorities. The Zoji La pass, which extends for several miles at an altitude of 11,000 feet, marks the boundary with Ladakh. This province, which is geographically a part of Tibet (its population is Tibetan in race and language, and Buddhist in religion), was under British rule. By myself I would never have been able to arrange this expedition, or to keep the costs within reason, but with the help of my companion Abid, everything went fairly smoothly. Our preparations included organizing a small caravan, including muleteers, a cook, and a few mules to carry our tents and provisions, for along the way we could expect to find little more than eggs and scrawny chickens. The cook, who had already worked for Europeans, showed himself to be clever and competent enough. Although I did not foresee major forays into the mountains, I yielded to the temptation of a mountaineer's ice axe that I happened to come across in Srinagar – a purchase that came in handy for a number of excursions off the beaten path. Its use, however limited, earned it a place among the other relics of my past that I still have at home.

The gorges of the Indus are supposed to be among the most beautiful in the world, and indeed I have never seen such an impressive sight. I will not attempt to describe them after an interval of fifty years. Our route was the one that leads to Central Asia and is followed by the caravans descending from there to Mecca. We met no other tourists except at Sonemarg, I believe, where the route opens up into a broad valley whose magnificent flowers in the month of June (we stopped there right in the middle of the month) have earned it its apt name (the "golden alp"). We did however cross paths from time to time with caravans drawn by yaks, the majestic, thick-fleeced bovines loaded with a variety of wares: woven and embroidered textiles, bricks of tea (which the Tibetans take with butter and salt), luscious dried apricots from the oases of Sin-Kiang, with a marvelous taste and smell I was not to come across again until forty years later, in Samarkand. These fruits, sold in shops in villages where we stopped overnight, formed a tasty addition to an unavoidably monotonous diet.

Such a journey would not be complete without some adventures. In the Zoji La pass, covered even in June with a thick layer of ice-encrusted snow, a caravan heading for Sin-kiang sent us a messenger who had great difficulty making himself understood, in bad Urdu, by my companion. The caravan was returning from Mecca. The previous winter, on its way west through the pass, the caravan had been caught in an avalanche in which men, beasts, and merchandise had been lost. At present, several meters of snow covered the whole area. Those who had survived had continued their journey, and were now on the return trip. They wished to recover the bodies of their companions, to provide them with a decent burial; perhaps they also wanted to get their hands on the precious wares that had been lost. But where were they to start digging? They had absolutely no idea, and, with a naïve faith in the miracles of western civilization, they counted on our providing them with binoculars enabling them to see under the snow. Alas, we could offer them no such help, and parted. The caravan drivers were as disappointed as we were dumbfounded.

In Ladakh itself, two days' walk from the Zoji La pass, just when we were about to turn back, we found ourselves equally incapable of responding to a more ordinary and yet more immediately tragic situation. In a peasant hut we were shown a man injured in an accident, his eye gravely wounded and already badly infected. His companions had improvised a makeshift bandage. For these people, as for many populations we refer to as "primitive," every European is at once doctor, surgeon, and miracle-worker. What could we say under the circumstances, except to recommend that the victim be taken to the nearest hospital? But Leh, the tiny capital of Ladakh, where there was no doubt an English or Hindu doctor, was several days' march from there; Srinagar was still farther. Fortunately, we did not have to dwell on this incident. We were invited to attend a polo match between two local teams. Not only is this game traditional in these regions, it is said to have originated there.

Not long afterwards, it was back to Aligarh, this time in the muggy heat of the rainy season. Vijayaraghavan had just arrived for the start of the academic year. As his name indicates, he was a brahmin from southern India. He came from one of those villages in the Tamil-speaking country where the traditional civilization of India survives in what is probably its purest form. His father had been a widely-respected *pandit*. In comparison with his father's, Vijayaraghavan said, his own knowledge of Sanskrit was poor. His modesty notwithstanding, he was thoroughly familiar with the ancient literature in both Sanskrit and Tamil. Like me with my pocket *Iliad,* which I had even taken to Kashmir, Vijayaraghavan never parted with a *Mahabharata* printed in Tamil characters, which took up two large grey cloth-bound volumes. Having failed his examinations as a young student in Madras, he had left to study with Hardy at Oxford, and had just returned to India when I met him. He was a very sharp mathematician, doubtless overly influenced by Hardy; but having no diploma, he hardly stood a chance of obtaining a post in any Indian university, much less in a Muslim university like Aligarh, but for the happy accident of my presence there. He had applied for the position as a long shot, and was quite astounded to find himself summoned for the interview and then selected for the position. It was not long before Vijayaraghavan and I were fast friends. With some exaggeration I may say that I never left his side. Even his mother, the reigning matriarch of the family, took me under her wing after observing on my first visit that I not only tolerated but relished an extremely spicy dish – which she had prepared, I am convinced, in the secret hope of scaring me away once and for all. I was the first European who had ever been admitted to her home. Even where her son's career was concerned, such a breach of the rules of caste must have been a source of some distress for her.

In one of the most beautiful stories of the *Chandogya Upanishad,* a young man named Satyakama ("lover of truth") seeks to become the disciple of a renowned master. According to the rule, he must be a brahmin by birth. Questioned on this point, he replies, without the least hesitation or discomfiture, that he knows nothing of his birth. His mother told him that he was conceived during a time of great activity in the house, and she does not know who his father is. Her name is Jabala and his is Satyakama. "I am therefore Satyakama Jabala," he concludes: that is all he knows. "Only a brahmin could speak so truthfully," the master responds, and he accepts the boy as his disciple. I like to imagine that my friend's mother, noting my taste for her hot curries, must have concluded likewise that I was a sort of brahmin that she had not yet come to know, or that I had been a brahmin in a previous life, born to the world this time in Europe as expiation for some sin, and that she could welcome me as such. In any case, she adopted me. She spoke only Tamil. I deeply regretted that I was not able to exchange more than a smile with her.

Naturally we conceived a plan to read Sanskrit together, but my friend's knowledge of it was too good, and mine too limited, for this activity to be very gratifying. On the other hand, Vijayaraghavan was an inveterate storyteller. At the slightest prompting, and even totally un-prompted, he would launch into tales from his beloved *Mahabharata,* or sometimes he would quote and comment on poems – gnomic, erotic, or mystic, in Sanskrit or in Tamil. Ancient Indian culture is one of the richest in the world. It ranges from the most abstract refinements of logic, gram-mar, and metaphysics, through the steamiest sensuality, to the purest mysticism. Vijayaraghavan took me beyond the initiation I had received at the hands of my Parisian masters: it is to him that I owe my true immersion in these cultural riches.

Later, as I have already mentioned, he presented me with a brah-manic cord. This took place in Dacca, where he also told me: "If you stayed in India, with customs changing as fast as they do, you would marry my daughter when she is old enough to be married." She was a charming girl, seven years old. He was joking – but, as always, only partly. Did he really think that a mere ten years would suffice to make such a marriage acceptable to his mother? Vijayaraghavan himself had been married in early childhood, and his wife had come to live with his family while still a young girl, long before they were old enough to consummate the marriage; he congratulated himself on this arrangement having been followed. Throughout his travels to Europe and America, I am positive that he was always absolutely faithful to his wife, just as he always remained a strict vegetarian. I think it was not so much that infidelity, or the act of eating eggs or meat, constituted a sin in his eyes; if he had, for example, eaten meat without knowing it, he would not have felt any remorse afterwards, and in fact he himself described such an incident to me. For him sin meant breaking a vow that he had made of his own free will. One day in Paris, when he had spoken of a woman with obvious admiration, my father asked him if he did not feel any "temptations." Vijayaraghavan replied: "I can enjoy looking at a Rolls-Royce, and take pleasure in imagining for an instant that it belongs to me, without being tempted to steal it." When, in the course of his travels, he felt the "prickings of the flesh," to use Rabelais's humorous expression, he would fast for two or three days. If only he had fasted more often! When I met him, he was already extremely corpulent. In the long run, his heart was unable to withstand the effort of moving such a weighty mass. His son described Vijayaraghavan's death to me: feeling the end was near, he asked to hear his favorite passages of the Vishnu liturgy, and these were the lines that accompanied him in his final moments.

Vijayaraghavan was thus my new recruit in 1930, and soon we began to draw up plans for the future of the department. I kept on traveling as much as I could. There were frequent holidays at the university, and I took full advantage of them. Naturally all Muslim holy days were observed. In addition, being under the British regime, and seeking to curry favor (in good Muslim tradition) with the authorities, the university zealously observed all the holidays of the British Raj: Christmas and Easter, as well as the royal birthdays. Moreover, despite its official designation, the university prided itself on not being sectarian: there were a small number of Hindu students, for the most part recruited locally, and a *pandit* to teach Sanskrit and Hindi. In fact a majority of the Aligarh townspeople were Hindu, and Muslim and Hindu coexisted on good terms. At least, I had the good fortune during my stay there not to witness any of the intercommunal strife which today remains a festering wound in Indian society, and which my Indian friends at the time commonly attributed to police provocation and refinements of British colonial politics. Whatever the case may be, courtesy toward the surrounding population dictated that the university observe the principal Hindu holidays as well. For me these interruptions provided welcome opportunities to travel.

The railway system was excellent. The master timetable for all Indian railways became my bedtime reading, and I still cannot think of it without waxing nostalgic. Distances are long in India, but even without airplanes they were not prohibitive. A night journey was nothing; a 24-hour trip was next to nothing. Everyone had portable bedding to lay out on his berth. There were four classes at the time: first, second, "intermediate," and third. First class was frequented by the English and by wealthy Westernized Indians who held important positions. In second class, which I preferred, I met mostly middle-class Indians. The less moneyed travelers who used the "intermediate" class knew little English; for this reason, I would travel in intermediate class only when I was with friends. Third class was for the masses, a teeming motley crowd that would fill the car to overflowing. I avoided it not out of class prejudice but first because of the physical discomfort and also because the linguistic barrier would have cut me off from my fellow travelers. It is true that with my friend Abid I had begun to study Urdu, a language that, though written in the Persian alphabet, is structurally identical to Hindi. I made some progress, but I never found the time to achieve fluency in speaking or reading. Besides, in many regions of India, it would have been of little use.

All over India, traditional hospitality was still the rule. The poor man, according to some classical text, is one who is unable to receive many guests in his home. One had to be very unlucky to have recourse to a hotel. When I told a friend that I was leaving for such-and-such a place, he would invariably ask me at whose house I intended to stay. If I said that I knew nobody there, I would be given an address. Sometimes, in the train, a fellow passenger would ask me the same question, and if need be would take me home with him. It is true that there was no need of a separate room for such a guest; a patio or porch with a cot was all the traveler needed to unroll his bedding. The bathroom was rarely equipped with running water. It was rare for an Indian to invite an Englishman to his home, and the latter would have been ill at ease had this happened. But to the Indians, I was not a colonizer. It is true that in many ways the French treated the Indochinese far worse than the English did the Indians, but the Indians knew nothing of this, and although I knew it I was not about to tell them. Before long I adopted the *achkan,* the "Nehru tunic," a practical wardrobe item worn by many colleagues and most of the students. I began by wearing one of black broadcloth but later, out of sympathy for the pro-Gandhi friends I soon made, I also had one in *khadi:* this was the coarse cloth that was hand-spun and hand-woven in keeping with Gandhi's teachings. I also wore the "Gandhi cap," made of the same fabric, which in France might have been called a police cap. From a distance, I could pass for Indian; sometimes, when the fancy struck me (ever since reading André Gide's *Les Caves du Vatican* quite some time before this, I had taken pleasure in passing myself off as someone I was not), I would introduce myself as a native of Kashmir. The Kashmiris are very light-skinned, and their dialect is so different from any other that outside their province they can speak only English. After a methodical apprenticeship, I had not only become accustomed to all types of Indian cooking, but I savored even the very hottest. I even enjoyed chewing the betel nut. My hosts, delighted to find me so enthusiastic about their cooking, were glad to provide me with food, shelter, and the pleasure of their company.

Thus my stay in India enabled me to visit it from one end to the other, from Kashmir to Bengal and from the Himalayas to Cape Comorin. It goes without saying that I visited museums and monuments everywhere. Among the places near to or far from Aligarh that I managed to see were Mathura, Jaipur, Madras, and Calcutta with their museums, Sanchi and its stupa, Akbar's tomb in Sikanderabad, Vijayanagar with its temples, and Delhi, to which I returned a number of times. But what purpose can such a random list of masterpieces serve, apart from rekindling my nostalgia? As for describing these places in glowing terms, guidebooks and art histories do more than enough of that. Suffice it to say that I was rarely disappointed. Need I add that India is no less rich in landscapes than in monuments? If I refrain from speaking of the former, it is not that my memories of them are any less dear to me.

In the course of these travels, I also discovered that India is not as poor in mathematicians as I had first feared. I met several in the north, and I saw a great many more in the south when, in 1930, the Indian Mathematics Society held its annual reunion in Trivandrum, the capital of the state of Travancore (today part of Kerala state). Along with my wish to meet mathematicians, I had a burning desire to get to know southern India. Two or three days of train travel no longer seemed a daunting prospect. I set off, stopping only in Madras. Although I would have liked to visit the great temples of the south – Thanjavur, Madurai, Kanchipuram, and many others – lack of time forced me to put off those visits to another occasion. Little did I know that I would have to wait over thirty years. In Madras I met Ananda Rau, a very gifted analyst, and Vaidyanathaswamy, also a friendly soul and a mathematician not lacking in talent, though rather limited in scope. In his company, I made the trip from Madras to Trivandrum.

At this meeting, I met only southern Indians. With very few exceptions, they were all brahmins, and recognizable as such by their names and the marks that many of them had painted on their foreheads. These signs designated them as belonging to one or another of the two large groups, known as *shaiva* (or *ayyar*) and *vaishnava* (or *ayyengar*) depending on whether they worshipped primarily Shiva or Vishnu. The president of the Mathematical Society that year was a brahmin of very high caste who came from Bangalore with his wife and a retinue of several women, both relatives and servants. He made it known immediately that he would not touch any food that was not prepared by these women's own hands. Caste rules dictate that a good Hindu must not eat anything prepared by a member of a caste that he considers inferior to his own, nor can he eat his meal in the company of anyone from such a caste. The man in question, who as president of the Society was moreover the personal guest of the Majarajah, intended to convey that no one, not even the cooks employed by the Maharajah expressly to serve the most distinguished guests, was qualified to prepare his food. Despite the pretense people made of laughing off this arrogance, it did not fail to make an impression. The president was a professor in a college in Bangalore. His brother, the director of an *ashram,* considered himself a divine incarnation, and I was shown some of his manifestos in which he served as interpreter for God speaking in the first person.

We were treated to a splendid excursion to Cape Comorin, which is at the very tip of the Indian peninsula. There my companions sang, not without emotion, the *Bande Mataram* ("I salute the Mother...") which was the Indian nationalists' anthem before it became the national anthem of independent India. We were also invited to a *kathakali* performance. This form of theatre acts out episodes of the *Mahabharata* in dance and mime. The ranking government official present was the Prime Minister of the State of Travancore, an imposing figure in his stately turban. But my clearest memory is of an incident that was both entertaining and revealing, and which deserves to be recounted here.

The members of the Society were housed in dormitories at the local college school, and took their meals in the school dining hall. More than in northern India, traditional customs continued to be observed in the south. For meals everyone sits cross-legged on the floor, or rather on an individual block of wood. The food, which in this case was strictly vegetarian, is served on banana leaves spread on the ground in front of each person. One eats with the right hand, without fork or knife, and drinks with the left hand; the hands are carefully washed before and after the meal. As the only European present, I was accorded the honor of being housed in a room in the "state guest house" generally reserved for English guests of the state of Travancore. This is where I was served dinner on the first night. Certainly English cooking deserves its terrible reputation, and what is known as English cooking in India is even worse. Moreover the prospect of eating alone was not a cheering one. I therefore told my friends that I wished to take my meals with them in the dining hall.

I knew that my request would present a serious problem. As a European, I was beneath their castes, or at the very least outside the system. The second day, in a little room adjoining the dining hall, they seated me at a table with Vaidyanathaswamy and one or two other colleagues, apparently used to associating with Europeans. We were served the same meal as the others. I feigned ignorance to the point of insisting on dining in the main hall itself for the next meal and was accomodated even to this extreme.

But the next day, I found that the rooms had been rearranged. It was explained to me that our group was too numerous for all of us to fit into the dining hall, so a second one was opened up; it was to this room that I was escorted. Most of the others in that room were young, and the ambience was merry. This arrangement lasted through the end of the meeting.

There was no point in inquiring about the matter, so I never did. Some of the elder colleagues, without perhaps protesting my presence directly, must have made the most of their right to observe the letter of tradition; if the younger ones wished to depart from it, that was their business. And thus a solution was found, to the apparent satisfaction of all.

At this meeting I was struck, as I had already been elsewhere, not by the level of mathematics – which was mediocre at best – but by the eagerness and openness of mind evident among the younger generations, a sharp contrast with the routine in which their elders were mired. I judged this to be a good omen for the future of mathematics in India. This optimism was actually somewhat premature, but later developments showed that it was not entirely unfounded.

My experience in Trivandrum also made me reflect upon the role of the southern brahmins in contemporary Indian science and society. Thanks to Vijayaraghavan, I had already had a glimpse of this role. It is hardly original to observe that for the last century or two in the development of Western culture, Jews have played a role that is far more significant than their actual numbers. This is particularly true in scientific fields, and first of all in mathematics. In India, southern brahmins have played an analogous role and continue to do so today. To cite only one example, when I visited the Tata Institute in Bombay in 1968, I could read the names on the doors of the offices on the floor reserved for mathematics: a large majority were those of southern brahmins, whose names are even more characteristic than the family names of Jews in Western nations. This analogy, of which I became aware in Trivandrum, calls for an explanation, which clearly is to be found neither in nepotism (though this has on occasion been operative in the success of both groups) nor in imaginary racial factors. And yet I do not think there is any great mystery to it.

To describe Hindu society, the West introduced the word "caste," which has been widely misunderstood. Indians prefer to speak of "communities". In this sense, a "community" is defined not on the basis of race or religion, but by specific patterns of behavior, including endogamous tendencies, culinary practices, and various rites. All of Indian society is thus compartmentalized in a system which includes everyone but saints, who by general consensus hold a place above the caste, as did Gandhi, the most celebrated modern example. Even a change of religion amounts only to leaving one community and joining another. All the reform movements that rejected the caste system (including Buddhism, long ago; the school of the philosopher Madhva in the middle ages; and in the last century, the *brahmo samaj*) simply created additional communities. It cannot be denied that this social compartmentalization has been somewhat relaxed during the past quarter-century, not least because of Gandhi's influence, and this relaxation is a good thing. Still, this system is too deeply rooted in the structure of Indian society for us seriously to envision its disappearance in the near future. In its essence it is not hierarchical, whatever some of our best sociologists may believe: the relative inferiority or superiority of castes is by no means inherent in this concept, and where such hierarchies exist they often rest on a subjective factor of which Indians are perfectly aware. Of two Hindus, it may well happen that each one regards himself as being of a better caste than the other.

Never to my knowledge have Western societies been so rigorously stratified; in any case, the concept of community cannot be applied today except to a few minorities, such as Jews for example, or Italian-Americans in the United States. Moreover, anyone is free to dissociate himself from these groups at will if circumstances allow, a possibility that does not exist in India. Such differences no doubt make the notion of caste, which is so clear in the eyes of the Hindus, so ambiguous and misunderstood among us.

Thus things become clear. Both the Jews and the brahmins of southern India are communities that, for twenty centuries, have devoted themselves tirelessly to the most abstract subtleties of grammar and theology. For the Jews it was the study of the Talmud, a task often passed down from father to son; for the brahmins, it was the *Brahmanas* and the *Upanishads*. It is hardly surprising that the younger generations, when their time came, turned toward the sciences, and preferably the most abstract among them: this trend was merely the natural extension of millennial traditions. Nor is it surprising that, in societies in which success is increasingly linked to certain intellectual qualities, both Jews and brahmins have excited the envy of others, the result being anti-Semitism in the West, and non-brahmin party and movement in India. For it does appear in our day that the southern brahmins have become the Indian homologues of the Jews in the West, and that hostility directed against them has taken some of the same forms (such as fixed quotas) that anti-Semitism has taken in Western societies. But it was not until some time later that these thoughts, the fruit of my experiences in India, occurred to me.

As to the academic year 1930-31, it was not without incidents. My teaching suffered both from my lack of experience and from the students' lack of preparation. Vijayaraghavan was the only one on whom I could count for assistance. The only English texts in the departmental library were completely out of date; I obtained some funds from Masood and entered into negotiations with bookstores in Leipzig. I appointed Kosambi for the following year. He was a young man with an original turn of mind, fresh from Harvard where he had begun to take an interest in differential geometry. I had met him in Benares (now Varanasi) where he had found a temporary position. I took it upon myself to institute a certain number of changes not only in the curriculum, but also in the examination system, which was, quite frankly, absurd. It was, however, the system that the English had established all over India, and my plan threw the students into confusion. I was still seen as Masood's creature, and a pamphlet in Urdu was circulated inveighing against him and criticizing me for importing French mathematics into India: according to the author of the pamphlet, only English mathematics was suited to the Indian mind.

Masood considered it quite a coup for Aligarh Muslim University that he had obtained from the Nizam of Hyderabad (under whom he had served as minister) funds for two richly endowed chairs, in physics and chemistry. To the chemistry chair he appointed a young Englishman whose merits I could not appreciate but whom I found friendly, and who became my neighbor. In physics, Masood thought it a triumph when he pushed through the appointment of a German whose only qualification was a letter of recommendation from Einstein, and whose merit in Einstein's eyes could only have been that he was an unemployed Jew – for he never displayed any other qualities. At that time, many European scholars thought that any European was good enough for a colonial country. To make this appointment official, procedure required that an *ad hoc* committee be convened with the participation of an outside specialist. The mistake was made of inviting the celebrated C.V. Raman, a Nobel laureate who himself had a student seeking a position at that very time. I believe this student was not without merit; in any case, he was without a doubt superior to Masood's candidate. I was also a member of the committee. Masood was in Europe, and cabled his orders. To contravene him would have touched off a serious crisis, and my fellow committee members, who were moreover utterly unqualified to judge the substance of the issue, would never have taken such a risk. Raman, justifiably outraged, threatened to cry foul and create a scandal. Vijayaraghavan joined forces with me after the committee meeting to dissuade him from doing so. An outcry would only have exacerbated an already highly uncomfortable situation.

In 1931, summer vacation did not begin until the month of May. I decided to spend it in Europe, not least in order to visit the bookstores in Leipzig to formalize the purchase of a library for my department. Of course I also stopped in Paris, Göttingen, and Berlin. Just when I arrived in Paris, Bruno Walter was directing the complete series of Mozart's great operas at the Théâtre des Champs-Elysées. I was fortunate enough to attend the performances, which were unforgettable. I had bought tickets for only one series of performances, but was so deeply moved that I wanted at the very least to hear *The Magic Flute* again, at all cost. Told there was not a single ticket left for sale, I put on my Nehru tunic and passed myself off as an Indian who had come to Paris for the express purpose of seeing this performance. I made such an impression that I obtained satisfaction.

Returning to Aligarh from Europe, I found the situation there to be rapidly deteriorating, for the university as a whole and for me in particular. All my efforts there had come to nothing – a failure which, while probably inevitable given the circumstances and my lack of experience, was no less a fiasco. Vijayaraghavan was no longer there. During my absence he had successfully applied for a position in Dacca, where he had already moved. I was shocked and dismayed. Later he told me that, after I had left for vacation, Masood had spoken to him, telling him that he, Masood, had plans to get rid of me, and offering him my position. Vijayaraghavan was so horrified by this deceitful move, which I of course did not suspect in the least, that he took the first opportunity he could to flee with all possible haste.

Nevertheless, I was still hatching more plans for the department. The books arrived from Leipzig, a rather nice collection of basic texts and periodicals, which I had chosen with great care as a foundation for serious scientific work. Thanks to Kosambi, I was not alone. I met Chowla in Delhi and made plans to hire him. In truth, my sole concern had been to gather together – it did not really matter where – a team of young mathematicians who truly loved their work. I believed that such a group would have a decisive effect on the future of mathematics in India. Perhaps my reasoning was sound, but to make this goal a reality I would have had to have more time, and first of all to have made myself unassailable.

In November, the university administration picked a trumped-up quarrel with me. I had neglected to seek official authorization to go to Allahabad to attend the inaugural meeting of a provincial academy. (As fate would have it, I was to be called on the carpet by a dean for a similar peccadillo years later, in Strasbourg.) But in Aligarh I had added insult to injury by refusing to help tabulate votes in an election held by the student club. The quarrel rapidly turned bitter. In January, while I was writing my letter of resignation, the university informed me that my contract was terminated. In my naïveté, I had never even asked to see this contract. No doubt the scenario was premeditated.

Virtually in a matter of days, I found myself without employment. It is true that I had informed some of my Parisian teachers, and first of all Sylvain Lévi, of my precarious situation. I was assured that a research grant, from what was to become the Centre National de la Recherche Scientifique, would be awaiting me upon my arrival in France. Furthermore, the salary Masood had granted me when I was recruited had turned out to be very generous indeed. Though I had been living with no thoughts of saving for the future, I was also without family obligations, and I found myself in possession of a sum which made me very comfortable for several months. At that time there were not even limitations on currency exchange to prevent me from spending it where I desired.

During the Christmas vacation, I had just visited Rajputana with Elie Faure, a physician who had become a rather highly regarded art historian. In his sixties, he had decided to take a trip around the world, and after visiting Japan he had, for reasons unknown to me, stopped in Aligarh. It is impossible to imagine a more agreeable traveling companion. He was constantly dipping into his unlimited store of tales, ranging from his memories of the impressionist painters to his amorous adventures and his trip to Japan. Describing a love affair in which he suffered the worst torments of jealousy, he told me, "If I had taken notes, I would have outdone Marcel Proust." By chance we arrived at Udaipur the day before the local prince ("His Highness the Maharana Sahib Bahadur," to use his offical title) had planned a panther hunt. Faure and I had no difficulty obtaining an invitation to accompany the prince. Perched together atop an elephant, we even used the shotguns that had – perhaps imprudently – been entrusted to us. I never found out if our efforts met with any success.

Once separated from Aligarh Muslim University, I did not miss it. I had some regrets that my efforts had failed, but at least they had succeeded in creating bonds of friendship among Vijayaraghavan, Kosambi, and Chowla, the three young mathematicians who seemed to me the most promising at the time. I also regretted having to leave India without having found the time to study Urdu and Persian seriously, as I had meant to do. In any case, I was in no hurry to return to France.

I made bold plans. One was to return by way of China and the Soviet Union. My sister, better informed than I about the situation in the Far East, did her best to discourage me. Another plan, which involved trips by plane, train, and even bus, was to visit Persia and Turkey. Sylvain Lévi suggested that I go to Japan, which held but little attraction for me: the extreme militarization which Elie Faure had described put me off.

Finally I adopted a more modest plan. Vijayaraghavan invited me to his house in Dacca, for as long as I wanted to stay. I decided to accept his invitation, and then to return to France by my usual Lloyd Triestino. On the way I stopped in Calcutta, which was not yet the collection of slums it has become since then, by all accounts. I remember not only the museum, which houses a deservedly famous collection of Bharhut sculptures, but also a magnificent exhibit of watercolors by Rabindranath Tagore (the poet, not his brother the painter). This exhibit gave me the idea of going to visit his Santiniketan foundation, an unconventional institution which combined instruction at all levels, from primary to higher education, in a rustic setting not far from Calcutta. A pupil of Sylvain Lévi's was teaching Tibetan there. He received me kindly and introduced me to the Poet, a majestic figure who held court in the park, dressed in sumptuous robes of purple silk and surrounded by deferential disciples.

Vijayaraghavan did not yet have his own house. He lived with his mother, his wife, and their darling little girl in a corner on the second floor of the main building of the university. I slept on a covered porch. In front of me was a tree with large red blossoms, whose name I didn't know (I have never known the names of trees, nor those of stars). I liked to think it was the same kind as the *ashoka* tree to which Damayanti, in the episode of the *Mahabharata* named for Nala and her, addresses such poetic speeches. My friend and I took our meals together, observing Madras custom: wearing the traditional *dhoti,* we sat on the floor and were served by his mother and his wife. The physicist Krishnan (Raman's student, who had shared, if not his teacher's Nobel prize, at least the work leading to the discovery for which the prize had been awarded) was a close friend of Vijayaraghavan's. We visited him among the rosebushes in his garden. Our group was completed by Shotyen Bose, the S.N. Bose of what is known as Einstein-Bose statistics.

Radhakrishnan, the philosopher who had just been named head of a newly-created university in Waltair, halfway between Calcutta and Madras, got wind of my presence in Dacca. He offered to make me head of his department of mathematics. I was looking forward to returning to France soon; nevertheless, after consulting with Vijayaraghavan, it seemed to me that, for the sake of mathematics in India, I did not have the right to refuse – provided that I be given *carte blanche* in choosing a team of mathematicians. I was then told that local politics made this condition impossible, and nothing came of this plan, save that Chowla was named to the post several years later.

These memories would not be complete without the inclusion, finally, of a friend no less dear to me than Vijayaraghavan: Zakir Husain. In Paris I had met his younger brother Yusuf, who was working on a doctorate in history. What he had told me about Zakir made me eager to meet the older brother, and shortly after arriving in Aligarh I sought him out. He was of Pathan ancestry, but had been born in a village of the United Provinces ("U.P.," known today as Uttar Pradesh). He had been a student at Aligarh Muslim University and was among those who, under Gandhi's influence, had left the university to found an experimental institution of which he became president. This was the Jamia Millia Islamia, more or less inspired by Tagore and his school at Santiniketan, of which it was in a sense the Muslim version. Like Tagore's foundation, Zakir's was neither school nor university. Its aim was to provide children and young people of all ages with a wholistic education consonant with Gandhi's principles of poverty, which in Europe would be called Franciscan. The Jamia Millia upheld the ideals and practices of Islam, without being a sectarian institution: there were Hindu students there, and Gandhi's son Deodas was teaching there when I visited.

At this time in India, it was beginning to be understood that an exclusively English view of the world was not suitable for India. Indeed, it was for precisely this reason that Masood had brought me to Aligarh. Zakir, I believe, was one of the first not only to understand this notion but to put it into practice. Before I knew him, he had spent several years studying in Berlin, where he had presented a doctoral thesis in economics. His German was almost as good as his Urdu, English, and Persian. He had sent his brother Yusuf to Paris for the same reason.

Zakir was already a legend when I came to Aligarh. Thanks to Yusuf, we quickly came in contact with each other. He immediately offered to be my host whenever I wished to go to Delhi, and indeed I was always his guest, even during my last visit: this was with my wife at the beginning of 1968. By then Zakir was President of the Indian Republic, and put us up in the former Viceroy's palace, the proud Viceregal Lodge (renamed *Rashtrapati Bhavan:* "the dwelling of the master of the empire"), with fully as cordial and as simple a welcome as earlier, when he had only a poor house without electricity in an outlying section of old Delhi, where he lived with his wife, their little girl, and the inevitable houseboy. I never saw his wife, who practiced purdah, the custom of reclusion observed by virtually all Muslims in India and, probably in imitation, by certain Hindus in the north. When I knew Zakir well enough to question him on the subject – that is, almost immediately – he told me: "That is the way she was raised; I do not try to influence her one way or another. I suppose she will always observe this custom, and that my daughter never will." His prediction came true.

This was a remarkable moment in the political life of the country. When I had arrived in India, Gandhi was on the verge of launching his campaign of civil disobedience through his famous march to the sea to protest the government's tax on salt. At Aligarh Muslim University, people made a show of not taking him seriously. While Masood and most of my colleagues were by no means Anglophiles, they were in any case loyalists, out of political conviction and tradition. They were not ready to comprehend that Gandhi was about to unleash a groundswell powerful enough to engulf all of India, simply by preparing a handful of salt in a small pot at the seaside. Gandhi himself knew full well. Setting aside all moral judgments, one might say that he and Hitler have been among the greatest publicity agents of all time.

Just as during the monsoon season Indian newspapers kept track of its progress from Cape Comorin to the Himalayas, likewise, during the spring of 1930, they followed the progress of Gandhi's march. At every turn the crowds swelled. Just as he had hoped, Gandhi was soon imprisoned, and the nationalist party, called the Indian National Congress, was declared illegal. Careful preparations had been made for this eventuality. The personnel destined to remain clandestine were hand-picked and limited to the minimum number possible. The party had branches in every village. As the known members of the local organizations were put in prison they would designate successors, who would join them there in turn. Some were motivated by selfless idealism and others, no doubt, by ambitions for future careers in politics. Gradually the prisons ran out of room for all those who were vying for the honor. The government was at a loss what to do. It must be admitted that, compared to what has been seen elsewhere, the repressive measures used could be described as benign. Someone told me, as an example of what he called "government atrocities," that for his medical examination upon entering prison he had been stripped naked. All I could think of was my appearance before the draft board in Rome, and I could not keep from laughing. Civil disobedience owed its fairly bloodless success to the fact that Gandhi was taking on not Hitler or Stalin, but the English. He never said this explicitly, to my knowledge, but he was far too much of a realist not to know it.

However close Zakir Husain was to the leaders of the movement, he had no formal ties to the party. For the duration of the campaign, he continued to devote himself to his tasks as an educator. It was for this reason that I was able to see him so often. I took advantage of my visits to him to glut myself on the glories of Old Delhi: the Red Fort, the Great Mosque (Jami Masjid, "Friday mosque"). Naturally, I was also eager to meet other members of Gandhi's prestigious inner circle, and most of all Gandhi himself. Thanks to Zakir and other friends, I met a number of these people over those two years, during the brief periods when they were not in jail. Furthermore, these people gladly made themselves accessible to all, insofar as this was possible. Thus it was that in Allahabad one day, Vijayaraghavan and I found ourselves sharing a small table with Gandhi. My friend introduced us to each other. Just then we were served tea – in coffee cups. Gandhi laughed. "It's easy to see you're not English," he said softly. "An Englishman would never have countenanced such a breach of etiquette."

On the same occasion, I made the acquaintance of the celebrated Rajagopalacharya, known as "the Gandhi of the South," who was to tell me in Madras 36 years later (by which time he was close to 90 years old): "I remember you perfectly; you have not improved since." I had already seen Jawaharlal Nehru in the autumn of 1930, in the resort town of Mussoorie, where I was recovering from a fever I had contracted in Aligarh at the end of the rainy season. He had been given a brief reprieve from prison for medical reasons, and was recuperating in his father Motilal's house. He explained to me that India would gain independence as soon as she began to cost the English more than the value of the revenues she brought: the goal of the movement was to hasten this moment.

Nearly all of the men who had been close to Gandhi and directly exposed to his influence seemed to have taken on some aspect of the personality of the Mahatma, whom they did not address as such, of course: to them he was Gandhiji, or better yet Bapuji, a term conveying filial affection. Zakir had adopted Gandhi's attitude of poverty and absolute devotion. He did not display the all-consuming need for activity which never seemed to allow Gandhi a moment's rest, whether from cleaning the latrines or bringing down the Indian government. For this, Zakir was too stately a figure. He readily referred to himself as *"Meine Kugelhaftigkeit (My Sphericity)"*. Where Gandhi was moved by an impassioned realism, Zakir evinced the perfect calm of an objective mind. Whether speaking of trivial incidents in his life or of the future of his country, he invariably spoke in the most everyday tone; his speech would never wax flowery or pretentious. I am sure this everyday tone is the one he must have taken in the celebrated interview that he conducted in Aligarh between Nehru and Jinnah at the time when preparations were being made to partition India. Zakir implored them not to let the situation go that far, and forced these two proud characters to embrace each other. That nothing more came of this interview was, I believe, one of the great disappointments of his life. I heard of this episode from other people, but when Zakir came to visit me years later in Chicago, I questioned him about the partition and the massacres surrounding it. I asked him whether his own life had been in danger. "In very grave danger," he told me, "and all because of my own stupidity." Having been ill, he had decided to go to Kashmir to recuperate. Without paying any heed to the date ("like a fool," he said), he had reserved a berth for the very night when the English government was to cease and desist, and India was to be cut in two. He left Delhi at the appointed time. In the middle of the night the train was stopped at the new border, where the Muslims were taken off and killed. On the other side, Hindus were being killed. No doubt it was impossible to tell where the massacre had started. A voice called to him: "Doctor Sahib" (this was how he was affectionately addressed at his Jamia Millia school), "what are you doing here? Come with me right away." He was taken to an office in the train station and locked inside. A sentry was stationed outside the door and told that he would pay with his own life if anything happened to the man under his guard. Zakir recounted this event as if narrating the most banal incident.

While in Aligarh, I used to read the *Hindustan Times,* the major daily newspaper in northern India. The press enjoyed a great deal of freedom under English law; who would have thought that forty years later, Nehru's daughter and successor would try to muzzle it? When I read one day that Gandhi was in Delhi to negotiate with the Viceroy in person, I took the first train and sped to Zakir's house. I found him just as curious as I was to be on the fringes of this historical event. He took me to the house of a wealthy physician, Doctor Ansari, who was a personal friend of Gandhi's, and the one with whom Gandhi always stayed when he came to Delhi. As usual on such occasions in India, the atmosphere resembled that of a carnival or pilgrimage. The negotiations lasted for days, in preparation for the London Round Table Conference. Every day, a Rolls-Royce was sent from the palace to pick up Gandhi; it would drop him off again at Dr. Ansari's in time for the evening prayer, the *gayatri,* which was sung by a faithful English disciple called Miraben. Especially at prayer time, all kinds of people crowded around to see Gandhi: even poor villagers would travel great distances to see him. In India, the sight *(darshan)* of a great man, especially one looked upon as a saint, has a religious value akin to that of the Pope's benediction for a throng of Catholics.

Miraben had a magnificent voice. According to a story that was told about her, she had once complained to Gandhi that her hair often provoked too much admiration from men. Gandhi answered: "If it bothers you so much, all you have to do is cut it." She shaved her head. It was Miraben who had assumed responsibility for preparing Bapuji's meals, a task that she would not have relinquished to anyone.

The day of his weekly vow of silence, Gandhi did not go to the Viceroy's. Once I found myself among the small group of followers who accompanied him on his walk that day. Many of them were out of breath from trying to keep up with him. Another day, it happened that the Viceroy (this was Lord Irwin, later to become Lord Halifax) wished to avoid interrupting the talks for dinner, and invited Gandhi to eat in the palace. Gandhi said, "That would not be possible; Miraben has already prepared my meal." The Viceroy suggested having the food prepared by Miraben brought to the palace, and Gandhi accepted. Thus, in this palace to which so many highly-placed Indians had considered it an honor to be invited even for a simple cup of tea, which no one would enter except in formal dress, where no one ate but from the finest china, Gandhi, dressed in his *dhoti,* ate his plate of lentils – the typical fare of the humblest Indian peasants – , served to him by the Viceroy's liveried servants. This incident had an obvious symbolic value and a propagandistic potential which the newspapers did not fail to highlight the following day. Masses of Indian peasants, eating their frugal dinner, knew from that time forward that Gandhi had partaken of the same meal in the Viceroy's palace.

On another occasion, I nearly witnessed an event – or perhaps more accurately, a hoax – which was probably calculated in part to cast ridicule upon the English administration. In this it succeeded. The very democratic constitution of the Indian National Congress granted powers to the central committee for one year only. These powers had to be renewed each year by a vote of the party congress, that is, the meeting of more than 600 local and provincial delegates who made up something like a parliament and from which the party derived its name. That year, despite the party's illegal status, the meeting was convened in Delhi. Naturally the police knew, but with the countless crowds entering and leaving Delhi every day by train or by car, on foot or bicycle, or in ox-drawn carts, it was impossible really to check everyone's identity, and only a hundred or so delegates were intercepted. In the evening, all the others were summoned in writing to meet at six in the morning on Chandni Chowk, the main artery in the center of old Delhi. The police found out, of course, and took action accordingly. Toward three o'clock in the morning, all the delegates were awakened by messengers with verbal orders to meet at four. At four o'clock, then, the two or three police officers patrolling the Chandni Chowk were overrun by the mass of delegates. The president, immediately designated by voice vote, climbed onto the roof of a car. Also by voice vote, a number of resolutions were passed, and the powers of the central committee were renewed. When the police arrived in force, the congress had already dispatched the day's business. Most of the delegates were taken to prison, as planned, and the Congress party had won the day. Arriving from Aligarh that morning, I heard almost all of these details from Zakir; the rest I got the next day from Âchârya Kripalani, who had masterminded the whole affair. He had chosen to remain underground for the time being, and I met him in the company of his fiancée's family, where he was in hiding – if it can be called hiding. If I had retained any of the grandiose notions I had acquired from Kipling about the efficiency of the Intelligence service, they would have been destroyed that day.

The English, on their side, were also seeking publicity, but less successfully. When the time came to name a successor to Lord Irwin in 1932, they thought they would perform a miracle by bringing the new Viceroy down from the sky – literally. As Boeings were yet to be invented, he was put in a dirigible, which went down in flames not far from Compiègne. I was in Dacca then. The newspapers announced the catastrophe in the most suitably tragic terms. All day, people would greet each other with a broad smile and such words as "What do you think of this tragedy? Dreadful, isn't it?"

What was happening with mathematics all this time? There was no idleness on that front either. I had reached an impasse with Diophantine equations. Hadamard's advice in such cases was to abandon the problem for a few years, in order to return to it later with a fresh mind. In his seminar, he often emphasized what was then called the "ergodic hypothesis." On this topic, he had never gone beyond Poincaré and Boltzmann. Even before leaving France, I had thought of applying von Neumann's recent work on unitary operators in Hilbert spaces to these problems. When I spoke to von Neumann about it in 1931, it seemed to me that the idea had not yet occurred to him, and he expressed interest in it. I thought it was a big step forward when I conjectured the truth of what is called the ergodic theorem in the L^2 sense. When I spoke to Elie Cartan about this, he objected that this finding was too general and imprecise to be really useful in the study of differential equations, and I ended up being convinced that he was right. I had a certain impulse to tackle celestial mechanics, which I knew Siegel was investigating; I soon gave up. In any case, I went back to Poincaré's celebrated theorem on the rotation number; I found a proof – an elegant one at that – that I hoped would make it possible to generalize the theorem to tori of more than two dimensions; but this intention did not pan out. I wanted in any case to extend it to all differential equations of the first order without singularities on the torus, and also, later, to the equations on compact surfaces of higher genus. For the first problem, which in fact had already been resolved by H. Kneser, I never worked out a satisfying proof. As for the second, I did not get very far at all. When I was back in Paris I managed to get my friend Magnier interested in it, but just when he began to get results, circumstances forced him to abandon his research.

I had more success with functions of several complex variables. I had been thinking about these for quite some time, and before leaving Paris I had had several conversations about them with Henri Cartan, which had renewed my interest in the subject; perhaps our discussions also had some influence on his work. French tradition taught that the theory of functions of one variable is dominated by the Cauchy integral; in truth, it is only one of a number of tools available, but I thought I had made significant progress by proving a formula that extended the Cauchy integral to very general "pseudoconvex" domains.

Every mathematician worthy of the name has experienced, if only rarely, the state of lucid exaltation in which one thought succeeds another as if miraculously, and in which the unconscious (however one interprets this word) seems to play a role. In a famous passage, Poincaré describes how he discovered Fuchsian functions in such a moment. About such states, Gauss is said to have remarked as follows: *"Procreare jucundum* (to conceive is a pleasure)"; he added, however, *"sed parturire molestum* (but to give birth is painful)." Unlike sexual pleasure, this feeling may last for hours at a time, even for days. Once you have experienced it, you are eager to repeat it but unable to do so at will, unless perhaps by dogged work which it seems to reward with its appearance. It is true that the pleasure experienced is not necessarily in proportion with the value of the discoveries with which it is associated.

I had known such moments in Göttingen in connection with Diophantine equations, but I wondered, and worried, whether they would ever return. When they did I was overjoyed. I was in Aligarh, and Vijayaraghavan was in Dacca. I sent him a wire saying: "New theory of functions of several complex variables born to-day," to which he jokingly replied: "Congratulations. Wire mother's health." I was certainly exaggerating; but perhaps I was not wholly mistaken in being happy with my discovery, which was similar to, but perhaps more complete than, the results Stefan Bergmann was obtaining at the same time. The first application I made was to a problem that had been posed some time before, on polynomial series. Oka, who was highly knowledgeable about this theory, to which he has made such valuable contributions, assured me much later that my finding had for some time played a virtually indispensable role. In any case, this result garnered me the most flattering praise I have received in my entire career. On my way home in May 1932, when I stopped in Rome to see Vito Volterra and explained my formula to him, he jumped up out of his chair and ran to the back of the apartment, crying to his wife: *"Virginia! Virginia! Il signor Weil ha dimostrato un gran bel teorema!"* ("Mr. Weil has proved a very beautiful theorem!")

Chapter V
Strasbourg and Bourbaki

While I was in India, my mother had kept me up to date on the eventful life led by my sister (the *trollesse,* or female troll, as we called her within the family) in Le Puy, where she had obtained her first appointment as a philosophy professor in a girls' *lycée.* Her activities during that year, and the following ones, have been described in colorful detail in Pétrement's biography; I will not duplicate that information here. Suffice it to say that because Simone had been so bold as to shake hands with unemployed workers in a public square, and then to accompany their delegation to the town council, the school administration had threatened her with disciplinary action. This news had reached me just when I was fighting my own battles with the administration at Aligarh, and naturally I was enchanted with the way my sister stood up to the authorities. I had written her a letter of congratulations, saluting her as "Amazing Phenomenon" – to which she replied addressing me philosophically as "noumenon." Le Puy was part of the Clermont school district. On my way back to Paris in May 1932, I stopped at Clermont, where I had friends, and took advantage of the opportunity to visit the rector. He seemed rather amused by my sister's exploits, which must indeed have furnished an entertaining diversion from the routine of his administrative functions. He told me that the climate in Le Puy was "not good for her, really not good." Knowing that by then she herself wished to be transferred elsewhere, I did not gainsay him, and we parted cordially.

Once I was back in Paris, I inquired about vacant positions in mathematics, and was told there was one in Marseilles. On the Boulevard Saint-Michel, I ran into the illustrious Denjoy, who took a friendly interest in my situation and asked me: "Young man, are you in the *cadres?*"[1] I had no idea that *cadres* were good for anything besides framing pictures; I must have appeared quite stupid to him. "But," he said to me, "it's very important for your retirement pension." This last word had little more meaning to me than *cadres,* and it was only much later that I understood just how pertinent his advice was. He left me saying: "Young man, think about your retirement."

---

1   *Translator's note:* The term *cadre,* besides being the French word for "(picture) frame," is used to refer to any one of the various levels of employment in the French civil service; *cadres* can also designate the employment roster of the civil service as a whole, including all teaching positions in public primary, secondary, and university-level education in France.

At the Ministry of Education, where I was assured of being appointed lecturer at the University of Marseilles, I was advised to write to my future dean. Pleased to learn that he was an alumnus of the Ecole Normale, I addressed him as *"Monsieur et cher Archicube."* I soon found out that he had been shocked by this form of address, and he had made no secret of his displeasure. He told me that he was planning to assign me the "general mathematics" course, which was the course for beginners, later known as "propaedeutic"; I do not know what it is called today. This assignment was hardly something to excite my enthusiasm. He urged me to be in Marseilles by the beginning of the academic year, and not to wait until my appointment had become official. "You wouldn't like to see your course deflowered by one of your colleagues," he wrote. I could not have cared less about such a defloration. At last my appointment was announced in the Ministry's official bulletin, to take effect as of December 1. I informed the dean that I would be in Marseilles on that date, and not before. He knew perfectly well that he could not demand more.

In the meantime, I spent the summer and fall traveling. After India, I was curious to see the English in their natural habitat. I therefore spent a few weeks in England, where I developed a lasting fondness for Cambridge and its colleges. I visited Mordell in Manchester; he welcomed me kindly and seemed flattered when I told him that without his 1922 article, I could not have written my thesis – about which, however, he showed no curiosity. This lack of interest was hardly surprising. Already in India, when C.V. Raman had asked me how many people had read my thesis, I had replied (thinking of Siegel and Artin) "one for sure, maybe two." Raman took pity on me for this lack of success, and was surprised to see that I was not disappointed by it. In Cambridge I was cordially received by R.E.A.C. Paley. A year younger than I, he was a very promising analyst. The following year he died in a skiing accident in the Rocky Mountains. Our conversation turned to a comparison of our approaches to work. At first, we seemed to be on completely different wavelengths. Finally, it became apparent to me that he worked fruitfully only when competing with others: having the rest of the pack at his side spurred him to greater efforts as he tried to surpass them. In contrast, my style was to seek out topics that I felt exposed me to no competition whatsoever, leaving me free to reflect undisturbed for years. No doubt every scientific discipline has room for such differences of temperament. What does it matter if a given researcher is motivated primarily by hopes of winning the Nobel prize? Sometimes it seems to me that Ganesh, the Hindu god of knowledge, chooses the bait, noble or vulgar, best suited to each of his followers.

I spent the rest of the summer in Switzerland, where the international congress of mathematicians was to take place in September, in

Zurich. The weather smiled on this event. There was a delightful excursion on the lake, which even a colleague bent on bombarding me with his latest findings did not spoil. Elie Cartan gave a memorable lecture, speaking in the same calm tone in which he had told me one day, as we walked down the Rue Gay-Lussac: "I am studying analysis situs, I think I may be able to get something out of it." A Picasso exhibition was on display in the Zurich museum. Rightly or wrongly, it seemed to me that his art juxtaposed the profoundly serious with the prankish – a mixture that was not without charm for the *Normalien* I still had in me. We mathematicians were only vaguely aware of what was called "the crisis." The congress was well attended, but not to such a point that one felt lost in the crowd – a feeling that has spoiled many another reunion for me since that time. Best of all, I was young. Zurich remains in my memory as the finest of all congresses I have attended.

On returning from Zurich, I still had several weeks of freedom, which I spent in Hamburg and Berlin. I thus had the opportunity, in Hamburg, to get to know Artin better, and in Berlin to attend a concert conducted by Bruno Walter and a performance of the Meistersingers. After this trip I had to get to Marseilles, where my duties were as easy as they were unappealing. Fortunately for my finances, the Centre National de la Recherche Scientifique (better known by its acronym, CNRS) was being inaugurated that year, with some confusion as to the role it was to play. It occurred to someone to use it as a means of granting supplementary income to university faculty members who showed some inclination toward what was just beginning to be called "research": up until then, this had been called "personal work." I was informed that I had been selected for such a grant. As a result, every three months for several years I received a check drawn on the national treasury, which I would cash in Paris on the rue de Rivoli. When people asked my sister what her brother did, she would answer: "He does research." "In what?" "How to get money from the government." The money I had brought back from India made me feel all the richer. I got used to traveling in sleeping-cars. Just as the master timetable for the Indian railroads had been my faithful bedside companion, now I was never without the *Mitropa* yearbook with the schedules of all major international trains. Thus it was that, finding myself in Weimar on a winter evening in 1933, I saw a poster for *Tristan* at the local theater, which had a good reputation. When I requested a seat at the box-office, I was told that there were very few seats left, and I had to take what I could get. People seemed to be staring at me intently, and I had no idea why. In the lobby during intermission, all eyes were turned in the same direction. When I asked, *"Was ist los?"* I was told: *"Das ist der Führer."* I did not attempt to get any closer to him. As I found out later, I had been present for the historical evening when Hitler, having just been naturalized as a "citizen" of the *Land* of Thüringen, was the guest of the Weimar aristocracy.

The city of Marseilles is not without its charms – at least, this was so until the war destroyed the old port. Nevertheless, I was not eager to stay there. Henri Cartan was already in Strasbourg, where a position was about to open up. We were both equally desirous of being there together. At the start of the academic year in November 1933, this wish became a reality. I taught there until 1939, with the exception of one semester spent in the United States in 1937. Those were happy and productive years.

Coming as I did from "the interior" (the term used by Alsatians at that time, and perhaps still today, for the rest of France), I was first of all struck by the appearance of the university's main building. This monument, like most of the surrounding neighborhood, is a typical product of the era of Wilhelm II. I told Cartan I was afraid I would not be able to get used to its unsightliness. He told me, "You'll see, one forgets it after a while" – but I never did get used to it. Fortunately the spacious and comfortable facilities of the mathematics department included a fine library, which was not only far superior to what was to be found at the time in provincial universities in "the interior," but also – not insignificantly – much more easily accessible to faculty and serious students. The excellence of this collection was due above all to the German mathematicians who had been there prior to 1918. Since then, however, fifteen years had passed, and it was no small accomplishment to have maintained and developed this excellent facility. The sole room for faculty was merely a sort of hallway off the library – there was no question of individual offices.

I was one of about ten young mathematicians from the Ecole Normale who, though scattered in various universities throughout the provinces as well as Paris, had maintained close ties since leaving the Ecole. My best friends among them, besides Henri Cartan, were Jean Delsarte and Claude Chevalley. Delsarte was a lecturer in Nancy, where he was to remain for his entire career which, unfortunately, was cut short by his premature death in 1968. Chevalley, back from Germany, was living in Paris. He had just been married and finished his thesis. Since Herbrand had died while mountain-climbing in July, 1931, Chevalley and I were the only ones left in France who were working on number theory. Each time I was in Paris, we met without fail.

I had spent quite a bit of time in German universities, and many of my friends had followed suit. There seminars played an essential role in education. The only seminar we had experienced up to that time in France was Hadamard's – a model we could hardly hope to imitate. We decided to organize one in Paris as a forum for regular meetings. At the time, such an enterprise required a "patron," if only to gain access to a room at the

Sorbonne. Julia, who had been the youngest of our teachers at the Ecole, was glad to assist us, so the seminar was known as the "Julia seminar." It continued until 1939. Unlike Hadamard's, our seminar focused on a different theme every year: in 1933-34, it was groups and algebras, then Hilbert space, then the work of Elie Cartan, etc. Mimeographed notes from the seminar may be found in the library of the Institut Henri Poincaré. Julia was a regular at this seminar, and no doubt this is where he got the idea of devoting the rest of his scientific career to Hilbert space. After the war, the "Julia seminar" was reborn from its ashes in a rather different form, known as the "Bourbaki seminar" – not that Bourbaki as such had a hand in it any more than Julia had had in "his" seminar earlier.

I went to Paris often, sometimes stopping in Nancy. In 1929, my family had moved into a sixth-floor apartment in a newly-constructed building at 3 Rue Auguste Comte, and for me they had taken a studio on the seventh floor with a view over all of Paris from the Luxembourg gardens to the Sacré Coeur church in Montmartre. From this distance the monument, silhouetted against the horizon, loses its monstrous character and almost becomes an essential element of an unparalleled urban landscape. I kept most of my library in this studio. My sister, pursuing her eventful career, had begun to receive her German friends fairly often at our parents' home. Our parents were sometimes troubled by these lengthy visits. Most of these Germans were dissident socialists or communists who had fled from Hitler. Unbeknownst to me at the time, even Trotsky stayed in my studio once, at the end of 1933. All of these comings and goings afforded me glimpses of the evolving political situation, without giving me any desire to get mixed up in it. At the University of Strasbourg, it was fairly common knowledge who was "on the left" and who was "on the right," and I was no doubt counted among the former. But political passions did not run strong there, and one of the colleagues with whom I got along best was a geologist who, though notoriously conservative, was also endowed with a healthy skepticism which I could not but share. In June 1934, during my last brief stay in Hitler's Germany (not counting a few excursions to the Black Forest), some young colleagues in Hamburg questioned me about rumors that had begun to circulate with regard to the horrors of the concentration camps. On that subject, I knew what I was talking about, and I informed them. Some of them refused to believe me, but one was easily convinced. Later, over coffee, he started making "subversive" jokes in a rather loud voice; his wife was frightened and told him, "Hans, not so loud!" He replied, "The reason I joined the S.A. was so that I could speak my mind." I met up with him after the war; nothing untoward had happened to him. 1934 was the year of the Night of the Long Knives, during which Hitler had some of his closest collaborators assassinated. The news reached me in a picturesque village in the Vosges mountains where I was correcting baccalaureate examinations. From that time on, I gave up traveling to Germany. One of the attractions of Strasbourg had been its proximity to the mathematics institutes in both Frankfurt and Nancy, but the former had been dismantled by the regime. Only Siegel lasted there for several more years, and his presence was invaluable to me. Several times we met for winter vacations in Switzerland or in the Black Forest, and it was in Arosa, at Christmas of 1935, that he told me of his discoveries on quadratic forms. As for my aunt and her family in Frankfurt, they knew enough to leave Germany in time. I was later to join them in America.

Satisfied though I was with my stay in Strasbourg, I had not lost my taste for traveling. One of the trips that marked me the most deeply was a journey to Spain in August and September of 1934, when I visited Toledo, the northern coast and Castille. I wisely put off until later a visit to Andalusia, which I finally made in 1936 at Easter. The familiarity I had acquired with Italian in 1925 gave me easy, if only superficial, access to Spanish – but the two languages have very different tonalities, and each seems uncannily suited to certain landscapes and certain states of mind. Is it possible to picture a bullfight in Italy? I did witness this game with death, at a time when no one had yet thought of filing the bull's horns to enable the *torero* to toy with the animal in a show of cheap heroics. The art of tauromachy was still at its zenith. In the Madrid arena, I saw the legendary Domingo Ortega perform a *corrida* that even I in my ignorance appreciated, and which was extolled in lavish detail in the next day's papers. The following year, Federico García Lorca was to publish his *Llanto por Ignacio Sánchez Mejías*. This elegy for an illustrious *torero* was to become Lorca's own requiem when he was assassinated by supporters of Franco in the first days of the civil war. Already in 1934, Lorca's great success was his *Romancero Gitano*, a slim volume of poetry pervaded by the idea of death: but then, isn't this the idea one finds at the bottom of the glass each time one drinks from the fountain of Spanish culture? True, the macabre *Dia de los Defuntos* rituals as practiced in Mexico are not found in Spain, to my knowledge; but wasn't the Spanish civil war one immense dance of death? *"Viva la muerte!,"* one of Franco's generals is said to have exclaimed after taking Salamanca. In 1934 I was struck by an incident described in the Oviedo newspapers. The political situation had already reached a feverish pitch. In response to rumors of an imminent arms shipment to a local extremist group, the police set up guardposts on the roads. During the night a car passed without hearing the orders to halt. The police fired shots, and someone was killed. Today, such an incident might be written off as an "administrative oversight." But the very same thing took place again the following night – without eliciting the slightest expression of surprise.

Among the books I brought back from Spain was a volume of Saint Theresa's writings. Being familiar with the *Bhagavad Gita,* and also to some degree with mystic Hindu poetry (the latter in translation), I was curious about a mode of thought that has always seemed foreign to me. I had also, unsuccessfully, looked for the works of Saint John of the Cross. The flashing beauty of his poems would probably have moved me more than did Saint Theresa, but it was not until much later that I came to know his work. I read a little of Saint Theresa and became quickly convinced that mystic thought is at bottom the same in all times and places: reading Suzuki's popular works on Zen was soon to confirm this conclusion. No doubt this is a widely accepted truth. If I mention it here, it is not that I am foolish enough to claim credit for it; but it hardly seems possible that my reflections on this subject were not somehow passed on to my sister, without perhaps her being fully aware of this influence. Perhaps we never spoke of it explicitly. But we knew each other so well that the slightest allusion – most often veiled in our habitual irony – was enough for us to understand each other.

My 1934 vacation ended with a visit to the Santo Domingo de Silos monastery near Burgos, to which I was attracted by a deservedly famous Romanesque cloister. The Benedictine monks, who were, I believe, affiliated with the Solesmes abbey, were most hospitable. From the conversation we had in the course of a ritual walk through the cloister, one sentence has stuck with me. Speaking of a saint whose behavior was somewhat eccentric, one of the monks remarked gently: "But Christianity is madness" *("el cristianismo es una locura").* This perfectly orthodox statement often comes to mind when I think about my sister's life.

Awaiting me upon my return to Strasbourg were Henri Cartan and the course on "differential and integral calculus," which was our joint responsibility – here I had escaped being saddled with "general mathematics." We were increasingly dissatisfied with the text traditionally used for the calculus course. Because Cartan was constantly asking me the best way to treat a given section of the curriculum, I eventually nicknamed him the Grand Inquisitor. Nor did I, for my part, fail to appeal to him for assistance. One point that concerned him was the degree to which we should generalize Stokes' formula in our teaching.

This formula is written as follows:

$$\int_{b(X)} \omega = \int_X d\omega,$$

where ω is a differential form, $d\omega$ its derivative, X its domain of integration, and $b(X)$ the boundary of X. There is nothing difficult about this if for example X is the infinitely differentiable image of an oriented sphere and if ω is a form with infinitely differentiable coefficients. Particular cases of this formula appear in classical treatises, but we were not content to make do with these.

In his book on invariant integrals, Elie Cartan, following Poincaré in emphasizing the importance of this formula, proposed to extend its domain of validity. Mathematically speaking, the question was of a depth that far exceeded what we were in a position to suspect. Not only did it bring into play the homology theory, along with de Rham's theorems, the importance of which was just becoming apparent; but this question is also what eventually opened the door to the theory of distributions and currents and also to that of sheaves. For the time being, however, the business at hand for Cartan and me was teaching our courses in Strasbourg. One winter day toward the end of 1934, I thought of a brilliant way of putting an end to my friend's persistent questioning. We had several friends who were responsible for teaching the same topics in various universities. "Why don't we get together and settle such matters once and for all, and you won't plague me with your questions any more?" Little did I know that at that moment Bourbaki was born.

If this is not just a synthetic memory, reconstructed retrospectively, if this conversation really did take place more or less as it is fixed in my memory, it would be of capital importance to Bourbaki's biographers to determine its date; but alas, I cannot. We all know that even the most distinct memories are not inscribed with dates on them, nor are they automatically arranged in proper temporal sequence. In many respects, memory is like a box full of old photographs or strips of film, some of them half blotted out; our various attempts to sift through them and reclassify them in chronological order are met with difficulty, and indeed we are often mistaken in our conclusions. Thus, as it is with many great historical figures, Bourbaki's precise date of birth will remain forever obscured. This suits him perfectly.

In any case, this conversation or one like it soon led to regular meetings including Cartan, Delsarte, Chevalley, Dieudonné, myself, and a few others, in a now-defunct Parisian restaurant on the Boulevard Saint-Michel. Those just named are the ones whose association with Bourbaki continued to the end, that is, until they reached fifty, their self-designated age of retirement. Later these few came to be called the "founding members."

There were Bourbaki archives, of which Delsarte took charge early on. For a long time, they were kept at the mathematics department at the University of Nancy; now they are in Paris. Unfortunately, they are rather incomplete with respect to the period under discussion here. What follows, therefore, is based primarily on my own memories of Bourbaki.

A series of legends has clustered around Bourbaki's name. Not least among their propagandists were his collaborators. The time has come to unveil these mysteries. As soon as plans for a collectively-authored publication took shape, it became apparent to us that we could not fill the entire cover with a long list of names. An old Ecole Normale prank came back to us at just the right moment.

When Delsarte, Cartan, and I were at the Ecole, the newly matriculated scientific class of 1923 received notification on the administration's official letterhead that a professor with a vaguely Scandinavian name would be giving a lecture on such-and-such a day, at such-and-such a time, and attendance was strongly recommended. The speaker was Raoul Husson, a more advanced student and a gentle prankster who later pursued a career as a statistician before finding his way to phonology and the scientific study of singing, to which he is said to have made some valuable contributions. In 1923, he appeared before the new "conscripts" armed with a false beard and an indefinable accent, and presented a talk which, taking off from a modicum of classical function theory, rose by imperceptible degrees to the most extravagant heights, ending with a "Bourbaki's theorem" which left the audience speechless with amazement. At least, thus goes the legend, with the additional detail that one of the students who attended the lecture claimed to have understood everything from beginning to end.

Our fellow *Normalien* had borrowed the name for his theorem from the general associated with Napoleon. When I was in India, I told this story to my friend Kosambi, who used it in a parodic note that he passed off as a serious contribution to the Proceedings of some provincial academy. Our group soon agreed to make Bourbaki the author of a future work. We had yet to specify which Bourbaki our author was to be. The question crystallized in 1935 when we resolved to establish Bourbaki's existence irrefutably by publishing a note under his name in the *Comptes-Rendus* of the French Academy of Sciences: we had to decide on a first name. My future wife Eveline, who was present at this discussion, became Bourbaki's godmother and baptised him Nicolas. We also had to get a member of the Academy to submit the note. We had no doubt that Emile Picard, permanent secretary to the Academy, would have an apoplectic fit if he got wind of this business. I volunteered to write the note and to send it to Elie Cartan with a supporting letter.

Elie Cartan was apprised of all our activities and plans. For him I made up a biography of Nicolas Bourbaki, describing him as of Poldevian descent. I emphasized that the Academy member who submits a note is obligated to ascertain the validity of its scientific content, but not to check the details of the author's biography. A group of members of the Academy was in the habit of meeting each week before the session for lunch, still referred to as the "young members' lunch"; if the truth be told, their youth was by then of rather ancient vintage. Elie Cartan was present at this lunch and, as liqueurs were being served, consulted his fellow members about my letter. He obtained their approval. As to the contents of the note, there was nothing fantastic about it – although someone later insisted that it contained an error: was this the inexorable effect of the Bourbaki curse?

Poldevia, Bourbaki's native country, was the product of another prank from the Ecole Normale. Around 1910, according to the story, several *Normaliens* made the rounds of cafés in the Montparnasse neighborhood and assembled a group of individuals of various nationalities, whom they converted (with the aid of a few drinks) into representatives of the Poldevian nation. The students wrote letters on behalf of these Poldevians, addressed to prominent political, literary, and academic figures, and beginning thus: "You are no doubt familiar with the misfortunes of the Poldevian nation..." Expressions of sympathy began to flood in. At an opportune moment, a public meeting was scheduled. The principal speaker gave a moving speech ending more or less as follows: "And thus I, the president of the Poldevian Parliament, live in exile, in such a state of poverty that I do not even own a pair of trousers." He climbed up on the table, and was indeed seen to be without trousers.

To conclude this digression on the name and origin of Bourbaki, I will add a more recent episode. Around 1948, Nicole Cartan called her husband to the telephone, saying "Bourbaki wants to speak to you." On the telephone Henri Cartan heard a voice saying, "My name is Bourbaki, and I would like to meet you." "I suppose you have a big white beard?" Cartan replied. (This was in fact how we liked to picture our creation.) "No, I do not have a beard, and I would like to meet you." Mystified, Cartan suggested a place for them to meet. At the appointed hour a distinguished-looking man showed up and immediately displayed a diplomatic passport issued in the name of Nicolaïdes-Bourbaki, an official with the Greek Embassy. He explained that his family was widely known. The Bourbakis counted among their ancestors two brothers who distinguished themselves in Crete in the seventeenth-century resistance against the Turks. On Napoleon's expedition to Egypt, his pilot was a Bourbaki. In gratitude, Napoleon granted this Bourbaki the privilege of having his son educated at the Prytanée de la Flèche, the national preparatory military school for sons of officers and government officials. The son became a French officer, and it was from him that Napoleon III's general, whose role in history is well known, was descended. Nicolaïdes-Bourbaki thought he was in possession of a complete family tree, and there were no mathematicians in this genealogy. How could it be, he wondered, that mathematical works were being published under this name? Cartan told him the whole story. From that time on, for several years, he was a regular guest at our end-of-congress dinners. In 1950, when I visited Greece, he gave me a letter for his relatives in Athens, where I was most graciously received. It is a pity that I was not able to make it to Crete as well, for I was told that a whole lamb would surely have been roasted for me in honor of Nicolas Bourbaki.

But an author alone was not enough: he also needed a publisher. The market for mathematical publications in France was dominated at the time by the Gauthier-Villars publishing house, which had carved out a near-monopoly for itself in this domain. But we had not the slightest temptation to seek their cooperation; they were far too academic for our taste. Fortunately, we did not even have to consider this solution. Right from the start we had a publisher: his name was Enrique Freymann.

He was as seductive a character as he was colorful. To hear him you would almost have thought that he was of pure Mexican blood: "I am an Aztec," he liked to say. As far as I could make out, he came from the state of Chihuahua, where one of his ancestors, a German forced to leave by the events of 1848, had settled and started a family. In everything Freymann said, whether about himself or any other subject, there was no use trying to separate fact from fiction, or rather, literal truth from essential truth. At a dinner to which Bourbaki had invited him, Freymann's wife begged him to tell a certain anecdote. He demurred: "that one isn't ready yet." Freymann said he had started out as a painter. After knocking about the world a bit, he had joined the Mexican diplomatic corps and married the granddaughter of the *archicube* Hermann, the founder of a scientific publishing house which, while never really rivalling Gauthier-Villars, had nevertheless published important works such as Elie Cartan's on invariant integrals in 1922. Freymann had taken charge of the publishing company, which he directed from the back of his little office on the Rue de la Sorbonne, with the help of two faithful employees and an errand-boy. In truth, Freymann *was* the Hermann publishing company. He hardly ever left his back office except to attend book auctions. In a dusty warehouse on the Boulevard Saint-German he amassed piles of rare books which he never really tried to sell. In 1929, he had started the series *Actualités Scientifiques et Industrielles,* a sort of spider's web by means of which he managed to lure the cream of the international scientific elite, as well as some of the dregs, into his den. For his series Freymann was open to any and all ideas, from the most thoroughly grounded in research to the most far-fetched, and he liked to say that some of the latter had been among the most successful in commercial terms. He gave little thought to the commercial aspects of the company, so long as he was keeping his head above water. Making up with one publication what he lost on another, he succeeded brilliantly, though I never knew whether to do so did not require miraculous feats of juggling. There were always kind souls to warn that he was on the verge of bankruptcy, although he never looked it. In his back office, he seemed to do nothing but shoot the breeze all day long. I would not even go there unless I had a few hours to spare, and I always regretted having to leave. Eventually Freymann and I became truly good friends. During the war, he saved my library by keeping it in his warehouse on the Boulevard Saint-Germain. His Mexican diplomatic passport, which he had kept, enabled him to get through those difficult years without mishap. In 1945 he himself told me how he had experienced the liberation of Paris. One morning, from up in his appartment on the Place de Medicis (the name has since been changed, alas, to Edmond Rostand), he heard shooting in the streets, and prudently decided to stay at home. The gunfire stopped and, as usual, curiosity got the better of him. He went down, made it as far as Saint-Germain-des-Prés, and saw that the neighborhood was already in the hands of the Free French. Already the newsstands were openly selling newspapers that had previously been forced underground. He bought *Libération* and *Combat,* put them in his pocket, and headed by way of the boulevard and the Rue des Ecoles toward the Rue de la Sorbonne. At a streetcorner he was stopped by a German patrol. The officer, gun in hand, frisked him and asked him in French: "Where did you get those newspapers?" Thinking he had nothing to lose by telling the truth, Freymann replied: "Don't you know they're being sold openly at Saint-Germain-des-Prés?" "How much did you pay for them?" "The same as they always cost, five francs apiece." The officer, still holding his gun, put the two newspapers in his pocket and took out ten francs, handing them to Freymann. "I am not a newspaper vendor!" cried Freymann, flabbergasted. "Yes, but *you* can buy them again, and I can't." Then he ordered his patrol: *"Vorwärts! Marsch!"*

In speaking of Freymann, I have fallen into his anecdotal mode and let myself get carried away. When Bourbaki began to acquire an identity, Freymann was already well acquainted with us. I suppose I had already met him before leaving for India. In 1931, several of my friends and I were deeply saddened by Jacques Herbrand's very recent death in a mountain accident, which left an irreparable void in our midst. We had decided to pay a final homage to him by publishing a collection of articles dedicated to his memory. Emmy Noether, von Neumann, and Hasse readily joined with us in this plan. Without hesitation Freymann agreed to publish this collection which, at his insistence, took the form not of a book but of a series of publications in his *Actualités* series. The minute he heard about Bourbaki, he was equally ready to assume the role of publisher. As it turned out, he had no cause to regret placing his trust in us and offering us his constant encouragement from the start. Bourbaki was destined to become one of the financial pillars of the Hermann company. But to have embraced our adventure at the time I am referring to redounds to his credit. There was no lack of priggish pedants to warn Freymann that he was making a fool of himself by getting involved in a vulgar college prank. Perhaps the Nicolas Bourbaki name and legend, the development and dissemination of which Freymann enthusiastically aided and abetted, were precisely what attracted him to our project.

The nature of our enterprise was not at first clear to us. In the very beginning, we had a more or less pedagogical aim: to trace the major lines of mathematical instruction at the university level. Soon we were talking about producing a text or treatise on analysis for use at this level, to replace the Goursat text as a basic curriculum. Our meetings in Paris were devoted to deciding on chapter topics and dividing up the work. Bourbaki asked his co-workers for reports on a large number of subjects, from set theory to analytic functions and partial differential equations. Gradually it became apparent that our Paris meetings were not sufficient for us to discuss these reports as fully as was called for, so we decided as a group to take two weeks of our summer vacation and spend them together in some place conducive to work. The University of Clermont had suitable facilities in Besse-en-Chandesse which were not in use during the summer. This is where the first Bourbaki congress was held, in July, 1935.

As unremarkable as this idea may appear today, it was far less so at that time. Somewhat later, Nazi mathematicians in Germany had the idea of organizing "work camps" on the model of the *Arbeitslager* where young unemployed Germans were sent to do manual labor. Since then, the institution has spread throughout the world, both capitalist and communist, and has become one of the most common ways of channeling government subsidies to scientific activities that are often quite meritorious. But even if they result in publications, these colloquia, conferences, symposia, congresses, or whatever they may be called are primarily designed as gatherings for mutual instruction. Such was never the case with the Bourbaki congresses, which were held, and in fact continue to be held, with the aim of collectively drafting and writing up a text. This is not to say that they do not afford an opportunity for the participants to learn from one another, but such exchange is not the object for which they were conceived.

As for outside funding, there was none at all until after the war. As we had chosen to meet for our own purposes, it seemed only natural that we foot the bill. After 1948, circumstances in France were such that a grant was requested from the Rockefeller Foundation. It was granted and gratefully received. Afterwards, royalties paid to Bourbaki were more than sufficient to cover our costs.

After the Bourbaki congress in Besse-en-Chandesse, I was called to a three-week military training session at the Mourmelon camp near Châlons. There, as a reserve lieutenant, I found a number of fellow *Normaliens* whose enthusiasm for military life matched my own. If I learned anything there, it was that this life consists primarily of waiting, and while the nights are short on sleep, the days on the other hand are mostly spent dozing. On the pretext of joining the group of French mathematicians invited, as I was, to the First International Topological Conference in Moscow, I obtained the Colonel's permission to leave the camp one or two days early. He asked me only one question: "Who is the commander of the detachment?" I had enough presence of mind to reply, "Monsieur Denjoy, a member of the Academy"; indeed, Denjoy was also among those invited. Permission was granted, so I was able to fly to Warsaw, where I got the train for Moscow without worrying about the "detachment." I arrived in time for the conference, which was scheduled for the week of September 4-10. Thanks to the generosity of my Russian colleagues, I extended my stay until October, including a brief visit to Leningrad.

This conference was the first mathematical meeting of such scope to be held in the U.S.S.R. (A Soviet conference held several years earlier in Kharkov, attended by a handful of foreigners, including Hadamard, did not really qualify as international.) It was not only the first but also the last such conference to take place under Stalin's regime. I myself was not much of a topologist, although I was not at all indifferent to the rapid evolution of topology. I owed my invitation, which I had received in April, to my friendship with Paul Alexandrov. I was too curious about the U.S.S.R. to do anything but accept it immediately. It enabled me to get a free visa, inscribed as such *("besplatno")*, in my passport. In Russia I was told that this was no small favor. In Moscow, foreign conference participants were put up in major hotels in the center of Moscow. We lived principally on the caviar canapes that were served in the hallways during the conference, for then as now, any attempt to have a meal in the hotel restaurants met with the passive resistance of the staff, which was virtually invincible. There is a modern Russian proverb to the effect that "the Russian people eat caviar by the organ of their best representatives"; apparently we were included among those representatives for the duration of the conference. Once it was over, I expressed a desire to prolong my visit by several weeks, and my colleagues in the Soviet Academy of Sciences kindly invited me to give a series of generously remunerated lectures. They put me up at the Scholars' House on a quay of the Moskva River, where I could also have simple meals without wasting too much time; Russian hospitality took care of the rest. I made some friends, among whom was Pontrjagin, who was after-

Bourbaki congress at Chançay (1936). Standing on bench: C. Chevalley's nephews; seated: A.W. and Chevalley's mother; standing, left to right: Ninette Ehresmann, R. de Possel, C. Chevalley, Jacqueline Chevalley, Mirlès ("guinea pig"), J. Delsarte, C. Ehresmann

wards to change so dramatically.[1] At that time he was young, light-hearted, receptive, bursting with ideas, and, as far as I could see, open-minded and independent-spirited. He was blind and lived with his elderly mother. I had already met Schnirelmann in Germany, and found him again in Moscow. He was an extremely talented mathematician whose premature death in 1938 prevented him from fulfilling his potential. Not until much later did I learn that he had committed suicide. He was a charming young man. The great misfortune of his life was that his lodgings consisted of no more than

1 *Translator's note:* Pontrjagin was to become one of the leading anti-Semites in Soviet mathematics.

a wretched furnished room, to which he was ashamed to bring his friends. It was with great embarrassment that he let me see it once. People told me that this alone was what had kept him from marrying.

Obviously I could get only the most superficial view of the situation in the U.S.S.R. I had gone there without the illusions to which a number of French intellectuals had fallen prey at that time. My sister had too much contact with dissident communist circles for me not to have some inkling of the true state of affairs. Among the "leftist" intellectuals, she was one of the first, not simply to see through these illusions to the true nature of the Stalinist regime, but also to perceive that the myth of the good Lenin as opposed to the bad Stalin was another illusion. Nevertheless, it seemed to me in 1935 that a certain optimism, which was not all for show, was shared by a number of Russian intellectuals, who seemed to believe that the worst of the oppression was over and that the regime was going to become progressively more liberal. The mathematicians, for example, could well believe that the support the authorities gave their conference was a symptom of this evolution. Perhaps I allowed myself to be persuaded. Alas, it did not take long for the great purges to open everyone's eyes – except for those who have not eyes to see; there will always be such people, as was seen in France at the time of the Algerian war and in the United States during the Vietnam era.

As for Russian mathematicians, they were for the most part unaffected by the purges. While in Moscow I was told an anecdote in regard to this. The Russian mathematician Otto Schmidt, whose name remains linked to a theorem of group theory, had been appointed to an important position in the government shortly after the October Revolution of 1917. At that time, it seems, he called together the principal mathematicians in Moscow and Petrograd (later known as Leningrad) and spoke to them more or less as follows: "Whatever the regime, the work of mathematicians is too inaccessible to laymen for us to be criticized from the outside; as long as we stick together, we will remain invulnerable." This same Otto Schmidt later became famous as one of the "heroes of the Tcheliouskine." His odyssey on an iceberg in the Arctic Ocean was the subject of a film, and acquainted the whole world with his noble face and majestic beard, which I had once admired in Göttingen. Schmidt outlived Stalin. I was told that in Stalin's time he felt safer in the far north than in the vicinity of Moscow.

I made another observation regarding the prevailing state of mind in Moscow in 1935. Since the October Revolution, the party line had included a strict internationalism. In 1935, Russians were once again permitted, or even encouraged, to adopt a patriotic stance. It seemed to me that they were reveling in this change. Furthermore, war was becoming a topic of conversation. When I visited Moscow's first splendid metro stations, which had just been opened to the public, I was told that they were also intended to serve as air raid shelters if necessary. When I expressed surprise at seeing the ubiquitous "closed for repairs" signs posted all over the place, especially in elevators, my friends explained that the best skilled workers were all in arms factories. Perhaps this was oversimplifying; but it seemed at least as plausible as invoking Dostoyevsky and the Slavic soul in this connection.

Having few occasions to hear comments on the regime itself, I found those I did hear all the more striking. My friend Zariski, who had become naturalized as an American and taught in Baltimore, was also in Moscow. He had a half-brother, an engineer in a factory in Siberia, who had stayed in the Soviet Union and who came to see him in Moscow. I was present for a good part of their exchange. This brother had taken part in the civil war and was now extremely bitter about what was happening in the U.S.S.R. What he said can be summed up in a few words: "It was not for this that we fought a revolution." The same conclusion has no doubt been reached by sincere revolutionaries in every century and every nation; but however predictable these words, how can they fail to touch? Zariski never had news of this brother again.

One day in the Tverskaia (later Maxim Gorky Street), I had a fortuitous encounter with a German auto mechanic. I had stopped him to ask directions in my bad Russian; he replied in German and we had an enjoyable conversation. We met several times in public parks in the city. He was the one who, on the day of my departure, came with a car to drive me to the station. At first I wondered if he were not an *agent provocateur.* But an *agent provocateur* is supposed to make others talk, whereas this man wanted nothing more than to talk himself, and he did so freely. He had left Germany out of disgust with Hitler's regime. "I might as well have stayed," he said; "it's no different here." Such a statement is hardly moving when it comes from a journalist or a public speaker. But in the mouth of a man who has lived what he is talking about, it takes on quite another value.

I also brought back from the Soviet Union a lesson of much more universal relevance. In the West, at the time I am describing, if one wanted to travel, one went to the station provided with the proper sum of money and bought a ticket. In the U.S.S.R., this method was only rarely practiced. Ordinarily, one sought a more or less plausible pretext to be issued a *kommandirovka* or travel warrant. The better one's standing with the authorities, the easier these were to obtain. I had already seen this system in action, on a smaller scale, at the Ecole Normale. By the time my sister was there, Lanson and Vessiot had been succeeded in the administration by Célestin Bouglé. The first two had tried in vain to keep the students docile by the use of ineffectual constraints. Bouglé, on the other hand,

exercised a far greater degree of control by shrewdly manipulating summer
travel grants and other petty favors. In the Soviet Union the same system
obtained on a much larger scale. The supreme favor was a *kommandirovka*
for international travel, and it remains so to this day. But why should this
be surprising? It is the rule everywhere; even prestige plays a role in this.
Does it not add to one's glory to fly off to Tokyo, deliver a more or less
abstruse lecture, and return the next day, all expenses paid?

Immediately following my trip to the Soviet Union was a colloquium in Geneva, in principle devoted to the same topic as the Moscow
conference, but conceived on a much more modest scale. Mathematically
speaking, however, it proved more fruitful to me. In Moscow, besides
Whitney's important exposé on fiber bundles on spheres, probably the
most original development – reported independently by both Alexander
and Kolmogorov – concerned cohomology rings of complexes and of
locally compact spaces. However, no doubt because I was too absorbed in
exploring my unfamiliar surroundings, I had not paid this topic the attention it deserved. In Geneva, where there were fewer distractions, the
lectures given by Elie Cartan and Georges de Rham did not fail to impress
me. Already, some years earlier, I had been struck by de Rham's application
of his theorems to algebraic geometry; in Geneva, I became fully convinced of the capital importance of these theorems, and of the notion of
"current" as de Rham presented it then. This was a provisional notion
anyway, for at the time Laurent Schwartz's "distributions" were not even
in gestation. Also in Geneva, cohomology rings made their appearance;
though viewed from a narrower perspective here than in Alexander's or
Kolmogorov's treatment, they were rendered more concrete by the use of
differential forms. These last were to become one of my favorite tools for
the study of varieties.

My own contribution to the Geneva colloquium had to do with
invariant measures on groups and homogeneous spaces. It was taken from
the volume on integration on groups which I had begun back in 1934 for
publication in the *Mémorial des Sciences Mathématiques,* to follow Elie
Cartan's famous contribution to this series. I resumed work on it in the fall
of 1935, continuing all the while my trips to Paris, for which the Julia
seminar (devoted that year to topology) and the Bourbaki meetings furnished ample opportunities.

In Strasbourg, Cartan and I were by no means idle. At that time, as
I had had occasion to observe in Marseilles, scientific life in provincial
French universities was virtually non-existent – but fortunately, Strasbourg
proved an exception to this rule. When the city was German, there had been
an excellent university: to give just one example, H. Weber had long taught
there. After 1918 the French wanted to maintain the university's prestige,

and for some years its faculty included eminent, even illustrious, scholars. Gradually, most of these – in sciences and humanities alike – succumbed to the charms of Paris; but still Strasbourg nurtured the admirable ambition of distinguishing itself from the grey monotony of the French provinces in general. This ambition was not confined to the university, but also made itself felt in the city's active musical life, of which I eagerly took advantage. Cartan and I were not as firmly resolved to put down permanent roots in Strasbourg as Delsarte was to establish himself in Nancy, but we felt no particular urge to leave. Our relationships with our senior colleagues – Thiry, Cerf, and Flamant – were excellent, and we found them always ready to encourage and aid our initiatives as best they could. I taught a course on algebraic number theory. I believe this was the first number theory course to be offered in a French university since the beginning of the century. In the university catalog, it was to be listed as "Arithmetic." To the dean, this title smacked of primary school; it did not correspond to his notion of the university's honor, so we changed the course title to "Number Theory" or perhaps "Higher Arithmetic," and he was happy. The dean was a specialist in earthquakes. I was not in his good graces because one day he had heard me ask a colleague, loud and clear, "Just what *is* a dean, anyway?" In 1934 he issued me an official reprimand for my having gone to Hamburg to give a lecture without obtaining official authorization (shades of Aligarh!...) Since the trip involved international travel, I ought to have asked for permission from the Ministry of Education, which would have consulted the Ministry of Foreign Affairs, which would have consulted the Embassy in Berlin, which would in turn have contacted the Consulate in Hamburg... This same dean once reproached my colleague Cerf for not wearing academic garb for some ceremony or other. "I will wear it," Cerf told him, "when the regulations on the wearing of academic gowns are properly enforced." "And what are these regulations?" "These regulations, which come from Napoleon and have never been rescinded, state that the gown shall be worn over traditional French attire, including knee-breeches, silk stockings, and sword." It came as something of a surprise to this dean when, not long afterwards, the Faculty elected another man to his position. The new dean was the astronomer Danjon, whose support for me in 1940 was instrumental in my obtaining a Vichy passport allowing me to go to the United States. He was to become director of the Paris Observatory after the war.

It has never been easy, particularly in mathematics, to attract to the provinces students who are concerned with more than the routine of competitive examinations; at the time I was in Strasbourg, it was nearly impossible. I did have two students whom I was able to direct towards "research." The first was Elisabeth Lutz, whose appetite for research had been whetted by my arithmetic course. Her thesis work on p-adic elliptic functions was accepted by Hasse for *Crelle's Journal.* Closer to the war, a very gifted student, Jacques Feldbau, asked me to suggest a topic in topology. I consulted Ehresmann, who was far better versed in the field than I. Following his advice, I suggested to Feldbau that he study the notion of fiber bundles, which was still quite young. Despite his somewhat clumsy methods (hardly surprising in a beginner) he came up with some interesting results, which appeared in the *Comptes-Rendus,* first under his own name and later, when the Vichy anti-Semitic laws made this practice inadvisable, under the name of Jacques Laboureur. He was deported by the Germans and died in a concentration camp.

But with Feldbau I am getting ahead of myself, far ahead. In the beginning of 1936, my future wife Eveline was in the process of divorce. At Eastertime I returned to Spain, this time with her, and we went as far as Andalusia. At the *feria* in Seville, we attended a magnificent *corrida,* for which I was careful to prepare my companion by making several stops on the way to the arena in Spanish bars where a delicious sherry known as *manzanilla* can be had. Thus primed, she had no trouble – nor did I, for that matter – sharing the enthusiasm displayed by the other spectators, who were after all much better judges than we in the matter of bullfights. The next day, butcher shops were advertising meat for sale from the previous day's contest; I suspect it must have been rather tough. On the way back to France, I was dazzled by the buildings of the Escorial (that concave sculpture against an immaculate azure sky) and did all I could to arrange for the summer Bourbaki meeting to be held in a high school near the monastery, which provided lodging for academic guests during the vacation period. How should I have foreseen the coming civil war? Many, more astute than I, were not expecting it either, or at least not so soon. It broke out in July; in August my sister left for Barcelona, and from there she went to the Aragon front. Our parents, justifiably worried, soon left in search of her. Not without difficulty, they managed to track her down to a hospital, rather the worse for wear. When I received the news, I replied with a postcard that infuriated my mother – or so she pretended for a long time: "Delighted to hear that you are all alive for the time being." No doubt I knew my sister was capable of the most rash improvidence, but what could I do about it? I had to prepare a report for Bourbaki on general topology. I wrote it in a hotel in a picturesque setting in the Eastern Pyrenees. For the most part, the outline I proposed was to be adopted the following month. Once this task was behind me, I devoted two weeks to touring Corsica, much of it on foot in the lovely forests in the northern part of the island – now laid waste, it appears, by recurrent fires. One evening I lost my way and happened upon the huts of some woodsmen from Sardinia. They treated me to a meal of polenta better than any I could have found in the finest Italian restaurant – and how much tastier it was in the middle of the woods, eaten at the very hearth where my hosts had prepared it! As I readied myself for the cot they kindly provided for me, I asked them what time they had to leave for work the next morning. "Whenever we want," they told me proudly: *"siamo i propri padroni* (we are our own masters)". In fact they set off at six o'clock every morning; but this was what they had decided of their own accord. Needless to say they refused to accept any payment for the warm reception they had given me.

In September, then, Bourbaki convened for the "Escorial congress," for this is what we called it even though the civil war prevented us from actually holding it there. At the last minute, Chevalley's mother offered to host us at their beautiful property in Chançay in Touraine, not far from Vouvray. Naturally, it did not take long for the work of the Bourbaki group to burst out of the overly narrow bounds we had originally set ourselves. The major classic texts in analysis (Jordan, Goursat) which we had at first set out to replace aimed to set forth in a few volumes everything a beginning mathematician should know before specializing. At the end of the nineteenth century, such a claim could still be made seriously; by now it had become absurd. Just to cover the most indispensable basics of set theory and algebraic and topological notions required significantly more than brief introductory chapters. When we undertook not to write a whole treatise on each of these subjects, but simply to treat it with enough breadth and depth so as not to have to keep adding to it constantly, it soon became apparent that there was no alternative but to give up any idea of writing a text for college-level instruction. Above all it was important to lay a foundation that was broad enough to support the essential core of modern mathematics; we might also dream about what could be built on top of this base, but that was not pressing.

At this congress and the next, our working methods took shape. For each topic, a writer was designated after a preliminary report and group discussion. This writer provided a first draft which the group would read and discuss again, modifying it to varying degrees or even, as happened more than once, rejecting it out of hand. Another writer would be designated to come up with a second draft, following the directives of the group – which of course were not always heeded; and so on.

Given this method, it was impossible to attribrute a given text to the name of any one member of the group. Moreover, it was agreed that decisions could be made only by unanimous consensus, and that any decision could be challenged; in cases of irreconcilable differences, the decision would be deferred. No doubt it required a major act of faith to think that this process would produce results, but we had faith in Bourbaki.

Nevertheless, we almost surprised ourselves when for the first time we approved a text as ready to go to press. This was the *Fascicule de Résultats* of set theory, adopted in its definitive form just before the war. A first text on this theory, prepared by Cartan, had been read at the "Escorial congress"; Cartan, who had been unable to attend, was informed by telegram of its rejection: *"Union intersection partie produit tu es démembré foutu Bourbaki* (Union intersection subset product you are dismembered f...ed Bourbaki)."* Wisely, we had decided to publish an installment establishing the system of notation for set theory, rather than wait for the detailed treatment that was to follow: it was high time to fix these notations once and for all, and indeed the ones we proposed, which introduced a number of modifications to the notations previously in use, met with general approval. Much later, my own part in these discussions earned me the respect of my daughter Nicolette, when she learned the symbol Ø for the empty set at school and I told her that I had been personally responsible for its adoption. The symbol came from the Norwegian alphabet, with which I alone among the Bourbaki group was familiar.

It was also at the "Escorial congress" that general guidelines for future publications – even the typography to be used – were established. To my great satisfaction (for the history of mathematics, or better said, the great mathematical texts of the past, had long fascinated me) we adopted the principle of following each chapter not only with exercises at various levels of difficulty, but also with a historical appendix; these "Historical Notes" were to contribute significantly to the distinctive quality of our work.

In establishing the tasks to be undertaken by Bourbaki, significant progress was made with the adoption of the notion of structure, and of the related notion of isomorphism. Retrospectively these two concepts seem ordinary and rather short on mathematical content, unless the notions of morphism and category are added. At the time of our early work these notions cast new light upon subjects which were still shrouded in confusion: even the meaning of the term "isomorphism" varied from one theory to another. That there were simple structures of group, of topological space, etc., and then also more complex structures, from rings to fields, had not to my knowledge been said by anyone before Bourbaki, and it was something that needed to be said. As for the choice of the word "structure," my memory fails me; but at that time, I believe, it had already entered the working vocabulary of linguists, a milieu with which I had maintained ties (in particular with Emile Benveniste). Perhaps there was more here than a mere coincidence.

I spent the rest of 1936 preparing for my trip to the United States, where von Neumann, with whom I had enjoyed friendly relations at least since 1930, had arranged for me to spend the second semester (from January through May, 1937) at the Institute for Advanced Study in Princeton. I had to put the finishing touches to the volume I had prepared for the *Mémorial*. Shortly before my departure I left the manuscript with Gauthier-Villars and arranged for Chevalley to replace me at Strasbourg during my absence. As was customary at the time, I would continue to receive my salary during this period, and it was up to me to remunerate him accordingly. We had no trouble coming to an agreement. I already had a subject ripe for a series of lectures in Princeton, though I still had to work out the details. This was the subject I had already addressed in my 1934 lecture in Hamburg, and which I was going to treat in the *Journal de Liouville* the following year. And so I made my way to Le Havre, where on January 10, 1937, I boarded the *Paris,* sailing for New York.

Fortunately I am not given to seasickness. A brief spell of it at the beginning of the crossing had no sequel, and there was nothing to spoil the pleasure I felt at the sight of the usual winter storms over the Atlantic. On stormy days the decks and dining rooms were nearly empty. The atmosphere on board was totally different from what I had grown accustomed to on the ships of the Lloyd Triestino that I had taken to and from India. This time there were so many passengers, and the crossing was so short, that we had fewer chances to become acquainted. I had been told that I would meet the Russian sculptor Ossip Zadkine, who was already famous. He had such a keen sense of color, it was said, that he would pack a tie to match his pale complexion on days when he was seasick. In any case, one calm day he was kind enough to open up his portfolios, which were full of impressive drawings. On the *Paris* there was also a South American music-hall artist, very proud of the "real cochon" act he had just performed at the Folies-Bergères with his partner. Two American aviators who were not sober for a single day of the entire crossing were permanent fixtures at the bar. One of these men had been among the first to fly over the Atlantic after Lindbergh. They were returning from Spain, where they had been serving the Republican cause – not out of conviction, but because the Republicans paid their mercenaries royally. Having served in a foreign army, they expected to be arrested the minute they disembarked – and sure enough they were. These two spoke fondly of the days they had recently spent in Paris, on the way to Le Havre. Their first order of business had been to retain the services of two prostitutes. These women, certain that the mercenaries' money would otherwise be squandered in no time, straightaway took the entire sum into their safekeeping, and when the time came to part returned what was left – after subtracting, of course, what was legitimately due them for their services.

Arriving in New York, I did not see the Statue of Liberty, which was hidden behind a dense fog. Like all travelers entering this monstrous city for the first time, I was at first dazed, stunned, overwhelmed. It took me numerous subsequent visits and a somewhat longer stay to become accustomed to it and feel somewhat familiar with it. It is true that at that time it was still a far cry from what it has now become: there were far fewer skyscrapers scraping the sky; there were far fewer madmen walking the streets. Central Park was peaceful and safe. Most saliently, racial hatred, while not totally absent, was at least less perceptible; I could stroll through Harlem, alone or with friends, and go to bars or movies there without exposing myself to verbal or physical aggression. Some of my most pleasurable evenings were spent in dance halls in Harlem, where one was always warmly greeted by the sound of jazz – sometimes very good jazz, according to friends who knew more about it than I. It was not uncommon for a black man to ask a white girl to dance.

As for the "culture shock" that awaits all newcomers to the United States, I had been fairly well inoculated against it by regular use of the American Library in Paris, which was then located on the Rue de l'Elysée. My "vaccine" consisted of reading not only Sinclair Lewis's novels, but especially books by H. L. Mencken, anthologies of his columns from the *Baltimore Sun* which were chock full of juicy as well as typically American anecdotes. I was therefore not terribly surprised by what I saw or what I read in the newspapers, where I enjoyed poring even over the classified ads – a regular practice of mine when in a foreign country. But I have nothing to say about all this that has not already been said a thousand times.

During my semester in Princeton, a visit to Harvard afforded me the opportunity to visit the Boston Museum of Fine Arts, which has a fine collection of Chinese painting from the Sung period. A young curator who insisted on guiding me through the exhibit assured me that only a knowledge of Zen would enable one to appreciate this painting fully. "So," I said to him, "you must have studied Zen yourself?" He confessed that he had tried, but without success. He and a few friends had sought out a Japanese monk living in New York and convinced him to come to Boston to initiate them. For the first session, they had invited a fairly large group of people to someone's home. The monk arrived at the appointed time, sat down in the lotus position, and announced, "Today we will meditate on the following subject," going on to state one of the classic Zen koans (perhaps it was "the sound of one hand clapping"): these are questions to which there is no possible rational answer; the object, I imagine, is to empty the mind to make room for enlightenment. The monk proceeded to meditate silently for an hour, after which he stood up, made a deep bow, and left. He was not asked back. I myself had gotten some idea (no doubt a very superficial one) of these methods from reading Suzuki on Zen, so to me there was nothing surprising about this anecdote. But, with or without Zen, the Chinese painting at the Boston museum left me with an unforgettable impression, for which nothing I had previously seen had prepared me, and which was reinforced but never eclipsed by the Chinese and Japanese collections I have since had the chance to admire.

The ambience in Princeton, which is still fairly cosmopolitan, was even more so in 1937. The Institute for Advanced Study did not yet have its own buildings; the University provided it with comfortable facilities in the old Fine Hall, to which Veblen had devoted so much care, but guests such as I were left to their own devices as far as housing went. Such stays are fruitful, but the experience has become such a common one that any remarks I could make would be superfluous. As planned, I gave a series of lectures on the topic of my future paper in the *Journal de Liouville,* and it was no small boost to my ego to see Hermann Weyl among those who attended regularly. Through contact with Alexander, I tried to find out more about "combinatorial topology," which was already among Bourbaki's interests. I met again with Courant, who invited me to visit him in New Rochelle, not far from New York, and who told me, "We still play music." To my query, "Quartets, as in Göttingen?" he replied, "No, now it's septets or octets *(Jetzt spielen wir Septett, Oktett)."* He was in America now: of course everything had to be "bigger and better" – at least bigger, if not better.

Everything comes to an end. In May, quite pleased with my stay in Princeton, but with no particular wish to come back, I took off for New Orleans, where I was to sail for Mexico. On the way I stopped in Baltimore to visit my friends the Zariskis, and then in Washington, where the mathematician Marshall Stone, who was approximately my age, insisted that I meet his father Harlan Stone, for whom his admiration (justifiably) knew no bounds. As a Supreme Court Justice, the elder Mr. Stone was a figure of considerable importance. My friend warned me that I should address his father as "Mr. Justice." I expected to be treated with formality, but the Justice – without any sacrifice of dignity – extended a simple and almost paternal welcome. I made bold to question him on the burning issue of the day. The Supreme Court had issued judgments that nullified some significant provisions of Roosevelt's New Deal legislation. The press spoke of nothing but Roosevelt's plan to nominate extra Supreme Court justices in order to create a majority favorable to his policies. Justice Stone explained to me that, even though he disapproved of it politically, he had voted for Roosevelt's legislation, since he could find nothing in it that contravened the Constitution, which was the Supreme Court's sole sphere of jurisdiction. The Justice had also tried to dissuade Roosevelt from this plan, counseling patience. In the end, Roosevelt relinquished it.

I spent a month as a tourist in Mexico, where I again met my friend Stone and his family and parents. In June, at Tampico, I embarked for England, where Eveline came to join me. Now that her divorce was official, we planned to be married in October. In the meantime, in September of 1937, Bourbaki was again to meet in Chançay. My sister, still intent on improving her knowledge of mathematics, attended our congress, and afterwards took several of us to the house of her friend Augustin Detoeuf in Combleux. This industrialist-cum-philosopher was the boss of Alsthom, where with his assistance she had become a factory worker – and where her health was irreversibly ruined.

Among the subjects under discussion by Bourbaki that year was general topology, as well as topological vector spaces. For several years, my penchant for Valéry, Rilke, and Claudel had led me to practice (to use Valéry's term) "the art of verse" (metered verse, free verse, eleven- and thirteen-syllable verse). To cite only a trifling example, the memory of a waterfall I saw in Kashmir had suggested a poem to me – a pale imitation of the poets I had read, but here nevertheless are the first lines:

> L'eau pure devant moi me blesse
> du remords de ne m'y plonger.
> Loin venus, lourds de ma paresse
> des souffles languides caressent
> mon corps, à moi-même étranger.[1]

In Chançay, Bourbaki inspired me with a sonnet, actually a fairly accurate summary of one of our discussions:

> Soit une multiplicité vectorielle.
> Un corps opère seul, abstrait, commutatif.
> Le dual reste loin, solitaire et plaintif,
> Cherchant l'isomorphie et la trouvant rebelle.
>
> Soudain, bilinéaire, a jailli l'étincelle
> D'où naît l'opérateur deux fois distributif.
> Dans les rets du produit tous les vecteurs captifs
> Vont célébrer sans fin la structure plus belle.
>
> Mais la base a troublé cet hymne aérien:
> Les vecteurs éperdus ont des coordonnées.
> Cartan ne sait que faire et n'y comprend plus rien.

---

1   The pure water before me wounds me with remorse because I do not immerse
     myself in it. From far off languid sighs, heavy with my laziness, caress my body,
     which is foreign to myself.

Et c'est la fin. Opérateurs, vecteurs, foutus.
Une matrice immonde expire. Le corps nu
Rentre en lui-même, au sein des lois qu'il s'est données.[1]

Back in France, a number of tasks awaited me. First of all, I had to find an apartment in Strasbourg for Eveline, her six-year-old son Alain, and me. Already the future seemed so shaky that we settled on a small furnished apartment on the bank of the Ill River. In spring the place turned out to be infested with bugs. When we told the landlady, she moaned and groaned about the preceding tenant, who, as she told us in her heavy Alsatian accent, had been the PO-lish CON-sul. We mostly ate lunch in a small tavern where we enjoyed the company of Dimitri Stremooukhoff and his companion Hela Chelminska, who became our friends. This exquisitely gracious man, the son of a Tsarist general, taught Russian, while she, also of noble ancestry, taught Polish. She was as forceful in her manner and speech as he was refined and delicate. When we met them they were unable to get married, for Hela would have lost her teaching position, which was sponsored by the Polish Embassy; Dimitri was stateless, had only a Nansen passport, and so could not hope to improve on his modest instructor's position, his remarkable doctoral dissertation notwithstanding. We were to find them again after the war, when he returned after having been deported. In a big raid, the Germans had rounded up virtually the entire faculty of the University of Strasbourg, who had taken refuge in Clermont-Ferrand. They had been sent to work in camps where they were so abominably treated that most, including Stremooukhoff, returned with their health ruined. Because of his internment he was naturalized as a citizen of France,

---

1   Suppose a multiplicity of vectors.
    A field operates alone, abstract, commutative.
    The dual remains aloof, solitary, plaintive,
    Seeking isomorphism and finding it defiant.

    Suddenly springs forth the spark, bilinear,
    From which the twice-distributive operator is born.
    All vectors, held captive in the product's nets,
    Will forever celebrate the finer structure.

    But the base has disturbed this airy hymn:
    The wild vectors have coordinates.
    Cartan is at a loss, no longer understands at all.

    This is the end. Operators, vectors, f...ed.
    An impure matrix fails. The naked field
    Withdraws into itself, within the laws it has decreed itself.

so that he was finally able to marry Hela and become a professor at the Sorbonne, before dying of cancer and exhaustion a few years later.

When I returned from the United States, my manuscript on integration in groups was lying dormant at the Gauthier-Villars publishing house – in fact, it was sleeping so soundly that it seemed nothing could wake it up. I retrieved it and took it to Freymann on October 30, 1937, which was also the day I got married. Delsarte, who had succeeded in attracting some of our fellow *Normaliens* to Nancy, and I took it upon ourselves to start an eastern branch of the *Société Mathématique de France*. Until the war, this branch held meetings in Nancy and Strasbourg alternately. At the most successful of these, Siegel gave a lecture on quadratic forms which was followed by a hike in the Haut Koenigsbourg. At the same time, I was writing up papers on several topics, including the material I had used for my course in Princeton. This paper was to be published, with a dedication to Hadamard (in words I had adapted from Lucretius), in the volume of the *Journal de Liouville* to be presented to him on the occasion of his retirement from the Collège de France.

Hadamard's retirement left his position open. I thought myself not unworthy of succeeding him; my friends, especially Cartan and Delsarte, encouraged me to be a candidate. It seemed to me that Lebesgue, who was the only mathematician left at the Collège de France, did not find my candidacy out of place. He even let me know that it was time to begin my "campaign visits." This custom was so well established at the Collège, as it was at the Academy and even elsewhere, that it would have appeared presumptuous for me not to observe it. I felt obliged to follow Lebesgue's advice. But at the same time, I became involved in a dispute which, to those few who still remember it, is known as the "war of the medals."

At that time, scientific life in France was dominated by two or three coteries of academicians, important persons, some of whom were visibly driven more by their appetite for power than by a devotion to science. This situation, along with the hecatomb of 1914-1918 which had slaughtered virtually an entire generation, had had a disastrous effect on the level of research in France. During my visits abroad, and particularly in the United States, my contact with many truly distinguished scholars had opened my eyes to the discouraging state of scientific scholarship in France. Upon returning from Princeton, I had written about this topic in an article entitled "Science in France," which I naïvely submitted to several periodicals. In it I discussed the boss system. The article was not deemed fit to print.

One of the cliques in question, and undoubtedly the most powerful one, was led by the physicist Jean Perrin, winner of the Nobel Prize, Undersecretary of State for Scientific Research and the inventor of the Centre National de la Recherche Scientifique. Not content with the significant powers he already held, he dreamed up plans to create a whole hierarchy of medals to be awarded with monetary prizes, ranging from the supreme gold medal down to lesser medalets worth 10,000 francs (a small sum at the time). The decree that established this system appeared in the newspapers during the Bourbaki congress in Chançay. It was not hard for us to guess that this scheme would be governed by the motto: *"Nul n' aura de l' esprit hors nous et nos amis."* [1] We were naïve enough to think that the joy of discovery was in itself ample reward. Worst of all, it seemed to us that this system would inevitably lead to further corruption in a milieu that we – relative beginners though we were – realized already had its share. We decided unanimously to circulate a petition in the universities, in the hopes that the Ministry would repeal the decree.

As a result, at the same time that I was recruiting signatures for the petition, I was also beginning my series of visits to the professors of the Collège de France in connection with my candidacy for the vacant position there. These professors considered it virtually a professional obligation to receive the candidates. I thus saw Paul Valéry, who had just been appointed to a chair in poetics. He asked me how old I was; I was 31. "Be sure to hang on to your age," he told me; "it's a prime number." I confess I was but little impressed by this pleasantry from a man whom I had admired without having met him. It is true that once, on a visit to Strasbourg, when a lady asked him to sign her copy of *Monsieur Teste* (a book beginning with the famous line "Stupidity is not my forte"), Valéry wrote: "Stupidity is becoming my forte, – Paul Valéry." I visited Marcel Mauss, whom I knew my Sanskrit teachers deeply admired. He was nearing retirement. He explained to me that he had long ago thought up a card system that would have enabled him to make important discoveries. "I was never given the necessary means when I could have made use of them; now that I've been given them, it's too late to do me any good."

In the course of my visits, I would solicit signatures for our anti-medal petition. This was hardly in the interests of my candidacy – particularly since it was said that the gold medal was to go to Hadamard. Finally Lebesgue put an end to my visits by telling me that he had decided on Mandelbrojt. It seemed to me that my friends were more disappointed than I at this outcome.

As for our petition, we had collected over 400 signatures: a rousing success in university circles. Four of us – Delsarte and I, along with a physicist and a linguist – presented the petition to the Minister of Education Jean Zay, who asked us in astonishment, "But do you understand that if

---

1 *Translator's note:* "But for our friends and us, none shall have wit": the quotation is from Molière's play *Les femmes savantes,* Act III, scene iii.

these two million francs are not allocated for medals, the University will
never see that money at all?" We chimed in unison, "Yes, Mr. Minister," a
response that only increased the Minister's surprise. It is all too widely
believed that it is better to misspend a sum of money than not to have use
of it at all. The Minister took no account of our petition. The endgame was
played during the final budget debates on New Year's Eve, with much
shuttling back and forth between the Chamber of Deputies and the Senate.
To preserve the fiction, by now long since outdated, of a budget actually
voted in before the new year, the clock used to be stopped at midnight in
both of these venerable institutions. In the Senate, Joseph Caillaux, the
indomitable enemy of all nonessential public spending, was poised with
his big axe, ready to chop the two million francs for the medals every time
Jean Perrin, in the Chamber, wrote them in again. It was only two weeks
later, after a thorough scrutiny of the *Journal Officiel,* that Delsarte and I
realized we had won. This victory was only a temporary one: the C.N.R.S.
has since reinvented the medal system, to the great satisfaction, it would
appear, of all concerned.

Chapter VI
The War and I: A Comic Opera in Six Acts

Prelude

Long before September of 1939, the shadow of the war had begun to stretch across Europe. Some, not without a certain fascination with the prospect, considered it inevitable; others did not think it impossible that the tempest would be averted at the last minute. My sister was among those who saw the impending war as the worst of all possible evils. Although she had never propagated or espoused the illusions of dyed-in-the-wool pacifists, she later came to repent bitterly for her original position. As for me, I tried my best to believe in a simplistic syllogism: "If England goes to war," I said, "then whatever happens, she will lose India; she will never accept the idea of losing India; therefore she will not go to war; therefore there will be no war." The prospects suggested by this reasoning were hardly more reassuring than the war itself, but they seemed in the long run to afford a bit more slack to humanity in this time of crisis, and I believe my sister shared this point of view.

Despite the recommendations of several friends in Strasbourg, I had not read *Mein Kampf*. If I had, I would have had even less faith in my syllogism, which had, besides, a major flaw: what seemed obvious to me – that is, that a war, even if it were won, would cause England to lose India – was not at all obvious to Churchill, who was always purblind when it came to India.

Like most of Bourbaki's collaborators, I was a reserve officer, liable to be drafted at the first call-up. In 1937 I had not yet decided on a plan of action in case of war. At Tampico, Mexico, I boarded a Dutch freight and passenger vessel bound for London; already then the Spanish war seemed pregnant with the imminent threat of a more general conflict. When the ship put in at Houston, I rushed out to buy newspapers, resolved to disembark if necessary in order to have time to reflect. This did not prove necessary, and I continued my voyage.

In 1938, Bourbaki held a congress in Dieulefit, where Chabauty, who had joined the ranks of the Master's collaborators, had family ties. Elie Cartan graciously joined us and took part in some of our discussions. This was precisely the time of the Munich conference.[1] There were sinister

---

1   *Translator's note:* The Munich Agreement between England, France, Italy, and Germany, approving the dismemberment of Czechoslovakia by Hitler.

forebodings in the air. We devoured the newspapers and huddled over the radio: this was one Bourbaki congress where hardly any real work was accomplished. By that time I had resolved that, if war broke out, I would refuse to serve. In the middle of the congress, after confiding in Delsarte, I thought up some pretext or other and left for Switzerland. But the immediate threat of war soon seemed to have dissipated, so I returned after two days. The day before everyone was to leave, Elie Cartan, who had to leave Dieulefit at the crack of dawn, shook hands with me, saying: "I probably won't see you again." Of course what he meant was "here in Dieulefit," but his words struck me as a very bad omen.

It seems fitting for me to indicate here the motives behind my resolution not to serve, though I fear this may turn into a long and confused explanation. I thought I was being perfectly lucid at the time; but is one ever perfectly lucid when making a decision of grave impact? I have never believed in the categorical imperative. The Kantian ethic, or what passes for it today, has always seemed to me to be the height of arrogance and folly. Claiming always to behave according to the precepts of universal maxims is either totally inept or totally hypocritical; one can always find a maxim to justify whatever behavior one chooses. I could not count the times (for example, when I tell people I never vote in elections) that I have heard the objection: "But if everyone were to behave like you..." – to which I usually reply that this possibility seems to me so implausible that I do not feel obligated to take it into account.

On the other hand, I had been deeply marked by Indian thought and by the spirit of the *Gita,* such as I felt capable of interpreting it. The law is not: "Thou shalt not kill," a precept which Judaism and Christianity have inscribed – to what avail?- in their commandments. The *Gita* begins with Arjuna, "filled with the deepest compassion," stopping his chariot between two armies, and ends with his lucid acceptance of Krishna's injunction to go to combat unflinchingly. Like everything else in this world, combat is an illusion: he does not kill, nor is he killed, whoever has known the Self.[1] In the absence of any universal recipe to prescribe everyone's behavior, the individual carries within him his own *dharma.* In the ideal society of the mythical times of the Mahabharata, the *dharma* comes from the individual's caste. Arjuna belongs to a caste of warriors, so his *dharma* is to go to combat. Krishna is the exceptional being, the divinity incarnate. "Whenever *dharma* declines and its opposite triumphs, then I reincarnate myself,"[2] he says in a famous verse which was once applied to Gandhi. Krishna exists outside the *dharma.*

---

1 *Bhagavad Gita,* II: 19.
2 *Bhagavad Gita,* IV: 7.

Bourbaki congress at Dieulefit (1938). Left to right: Simone Weil, C. Pisot, A.W.; J. Dieudonné (seated); C. Chabauty, C. Ehresmann, J. Delsarte.

It is true that Arjuna describes to Krishna in advance what must result from the deadly combat that is about to take place: not only the death of loved ones, but also social chaos, the fall of woman, the confusion of castes. Krishna does not respond to this description; but the battle of Kurukshetra ends with all of humanity nearly wiped out: it was not for nothing that the ultimate weapon was in Arjuna's hands. A curiously modern dénouement indeed! All that Krishna, incarnated, brought to the world was the *Gita*. It is no trifle.

If we take these teachings seriously, what are we to conclude? Clearly we now live in the total confusion of castes; the only recourse is for each one of us to determine as best he can his *dharma,* which is his alone. Gauguin's *dharma* was painting. Mine, as I saw it in 1938, seemed clear to me: it was to devote myself to mathematics as much as I was able. The sin would have been to let myself be diverted from it.

I was not unfamiliar with the *Criton* and the prosopopoeia of laws. But between the supreme obedience Socrates pays to the laws of his city and Gandhi's civil disobedience, I did not feel that the former had to be preferred to the latter. Gandhi did not justify his conduct by the fact that the laws he chose to disobey had been imposed by a foreign government, but by the sole fact that they were unjust. Moreover, for him (contrary to popular belief) it was not a matter of the *right* to disobey unjust laws, but rather of the *duty* to disobey them – another thing altogether. According to Gandhi, duty meant disobeying laws whenever one is convinced that they are fundamentally unjust, regardless of the consequences of such disobedience.

Already while at the Ecole Normale, I had been deeply struck by the damage wreaked upon mathematics in France by World War I. This war had created a vacuum that my own and subsequent generations were hard pressed to fill. In 1914, the Germans had wisely sought to spare the cream of their young scientific elite and, to a large extent, these people had been sheltered. In France a misguided notion of equality in the face of sacrifice – no doubt praiseworthy in intent – had led to the opposite policy, whose disastrous consequences can be read, for example, on the monument to the dead of the Ecole Normale. Those were cruel losses; but there was more besides. Four or more years of military life, whether close to death or far away from it – but in any case far from science –, are not good preparation for resuming the scientific life: very few of those who survived returned to science with the keenness they had felt for it. This was a fate that I thought it my duty, or rather my *dharma,* to avoid. So I never seriously considered the solution proposed by my brother-in-law, a career officer: he suggested using my extreme nearsightedness as grounds for transfer from the infantry to the Quartermaster Corps. Besides, remembering how scornfully people had referred to the *"embusqués"* (the slang word for shirkers, who had managed to avoid seeing action in the previous war) was enough to dissuade me from this solution. I was not a conscientious objector in the sense usually given to this word, that is, someone who does not believe in killing or even using instruments of death, whether he sees this as a universally valid precept or as his own personal *dharma*. It has always seemed to me that, logically, priests and monks ought to be among those for whom this belief is an indispensable way of life; I have never understood by what theological subtleties they allow themselves to be made into soldiers. As for myself, it is true that I find killing repugnant, even if it is only a fly that is bothering me; but if I am unable to shoo it out the window, inflicting a violent end upon it does not fill me with remorse. For the same reason, the issue of abolishing the death penalty seems to me simply a practical question, which I do not feel capable of resolving; and if I saw a man, whatever side he might be on, set fire to the cathedral of Chartres, and I had a weapon in my hands, I would not for a second hesitate to shoot down the arsonist rather than let him go about his business. In other words, I feel myself as far removed from unconditional pacifists as from intransigent patriots, if any of these remain, or from fanatical leftists. In 1940, when the line of conduct I had adopted brought me a series of problems, my sister was apparently stricken with remorse at the thought that her pre-war pacifist views could have influenced me. She was mistaken in this conclusion; besides, as I have already said, her pacifisim, unlike that of some of my friends, was not unconditional either, but above all pragmatic and, she felt, realistic – even though she was later to change her mind on this question.

If any example had influenced me, it was Siegel's. One day, in a hotel in Switzerland, he told me how he had deserted in 1918. Drafted as a young student, he had just begun to serve in Alsace when he decided that this war was not his war (*"Dieser Krieg war nicht mein Krieg,"* he said); he took off, disappearing into thin air near the Haut Koenigsbourg, where he gave himself over to unhappy reflections. Naturally he was caught. In order to spare him a court-martial, they sent him to a psychiatric hospital. He never spoke to me of his internment there, but for many years he bore the mark of this experience. Siegel's story held a certain beauty for me; it seemed to me that in certain circumstances, such conduct can have an exemplary value. Did I flatter myself that mine could as well? Quite possibly: what corner of the human heart is immune to vanity?

How could a Frenchman, in 1938 or 1939, have accepted this imminent war as his war? The men who led him into it and insisted on their right to do so were the very same people whose incompetence and blindness had both made the war inevitable and compromised its outcome in advance. For better or for worse, one does not lead a country into war without some qualities of leadership. The French at this time behaved more like sheep resigned to follow wherever they were led – be it to the slaughterhouse – than like free men resolved to defend what is most dear to them. In this light, I felt justified in thinking that by exempting myself from military law, I was, to the small extent permitted me by circumstances, taking charge of my own destiny.

Naturally, nothing happened according to plan. I had imagined that the coming war would be much more like the previous one than it actually turned out to be. This illusion was shared by many apparently clearsighted minds, as well as by the great majority of the French general staff, who were anything but clearsighted. As to the aspect that concerned me particularly, the war proved much less of an interruption to the work of the Bourbaki group, and that of other French mathematicians of my generation and the following one, than I feared it would be. Would I have taken the same position in 1939 if I had known what was to come? An idle query: for better or for worse, I had made up my mind.

My plan was, in case of war, to take refuge in a neutral country, and then to emigrate to the United States. I did not know then that the Americans, who so warmly welcome those who do not need them, are much less hospitable to those who happen to be at their mercy; no doubt this is a universal human trait. At the time of the Munich accords, I had fled to the nearest refuge, that is, to Switzerland. After the Bourbaki congress, the clouds did not appear to be dispelled, so I went to the Netherlands to await further developments. When Neville Chamberlain announced "peace in our time" (and, of course, an "honorable" peace – as did Nixon toward the end of the Vietnam war), I decided that I could return to Strasbourg, with a jog over to London on the way. There I was shocked to learn that the negotiators in Munich had not received any substantial guarantees from Hitler with respect to the future. Friends in London assured me that the showdown had merely been postponed. The librarian of the Royal Geographical Society told me that the previous summer, in the thick of the Czechoslovakian crisis, she had received an urgent request from the Foreign Office for any information she could possibly send them in relation to Czechoslovakia. She had not found this reassuring as to the competence of the leaders of her country.

Returning to Strasbourg, I found that the Avenue de la Liberté had been renamed the Avenue Edouard Daladier;[1] the Avenue de la Paix was now named for Neville Chamberlain. These changes struck me as charged with a sinister symbolism. Commenting on the Munich Agreement, Henri Cartan said: "It's like vomiting: you feel sick but also relieved." I continued teaching my courses, all the while letting my American friends know that I would be gratified to be offered a university position in their country. This

---

1 *Translator's note:* Daladier was by then both Prime Minister of France and Minister of Defense.

solution would have greatly simplified the next phase of my life, but my suggestions led nowhere.

Quite early one spring morning in 1939, I received a telegram from my sister: "Recommend read newspapers." Hitler had just taken over Czechoslovakia. I was immediately convinced that France would not budge; nor did I. Soon the following joke was being told in England, at the expense of Neville Chamberlain, who was supposed to have said: "In Bad Godesberg I realized that Hitler was a bandit, and in Berchtesgaden I realized he was crazy; but now, after the invasion of Czechoslovakia, I know he is no gentleman." France had mobilized a few so-called "protective" units. Shortly after these events, when I was on the train to Paris, I heard a reservist from Nancy describing, amid peals of laughter, how his regiment had left its quarters in Nancy to occupy a fortification on the Maginot Line: when they had reached their destination, it turned out that the officer in charge had forgotten the keys to the casemates. Someone had to be dispatched by motorcycle to fetch them with all possible speed.

Finnish Fugue

My wife and I were friends with Lars Ahlfors. In 1939, it was arranged that we would spend a few weeks of the summer with him and his family in a villa he was planning to rent on the Gulf of Finland. If war did not break out by summer's end, we would return to Strasbourg, perhaps with a detour via Leningrad. Otherwise, my plan was to remain in Finland, where I thought I would have ample time to prepare my voyage to the United States. This was not my only miscalculation. I took with me, in dollars, what I thought would be enough money to cover my expenses. Eveline, who of course knew of my intentions, thought no good would come of these plans.

Our visit with the Ahlfors was a time of unadulterated serenity. The rented villa was on a tiny island called Lökö, ("Onion Island"), a half-hour's row from a larger, busier island, Lille Pellinge ("Little Pellinge"), which was connected to Helsinki by the boat *Lovisa*. On Lille Pellinge there was a small tourist center where we went for supplies. Our small island was easily explored: besides our villa, there was only a small farm with four or five cows, some distance away. It was the season of white nights, close to midsummer night's eve. The air was invariably pure and clear, transparent beyond words. As the possibilities for walking were extremely limited, we went instead on long outings by boat, exploring the rocky islets nearby. The Baltic waters were so cold that we went for only brief dips. With Lars Ahlfors was his wife Erna, their daughter Cynthia who was beginning to stammer her first words, and a beautiful dog. Erna was pregnant with their second daughter. At around ten or eleven in the evening, we would gather in the dining room for tea and sandwiches by the light of the setting sun. We never lit a lamp; probably there wasn't even one on hand. I do not recall whether we received newspapers there. The passing of the days was marked by the regular appearance of the *Lovisa* off the coast of our island. We felt we were somewhere outside of time.

After this visit, Eveline and I were invited to spend several days with Nevanlinna at his lovely country home on the shores of a lake not far from Helsinki. He was there with his wife (from whom he was later to be divorced) and their darling daughter Silvia, who immediately took a liking to Eveline even though they could not exchange a word. The Nevanlinnas' hospitality could not have been more cordial. Mrs. Nevanlinna was a great admirer of Hitler, and it seemed to me that her husband's sentiments were similar. To judge by the welcome they gave us, however, these views were not tainted with any anti-Semitism. For them, as for many Finns, fervent patriotism was inseparable from hatred for Russia, which was itself insep- arable from hatred of communism. To them Hitler appeared as the future savior of Europe, and I think they were willing (too easily, no doubt) to close their eyes to what must to them have seemed excessive in his regime. Among Mrs. Nevanlinna's books I found *Mein Kampf,* which I read in one sitting, and which left me with no doubt as to what was to ensue.

Our next stop was at a small hotel on the shore of Lake Salla, quite close to the Russian border, where there was to be heavy fighting the following winter. Our days were calm and uneventful. We spent many hours on the lake in the hotel skiff, or sitting at the water's edge. I had with me my faithful typewriter, and there I typed the outline for a report on integration that I had promised Bourbaki. As for Eveline, she had her little shorthand typewriter with her. Before our marriage, she had trained as a stenotypist. If the need were ever to arise, she could support herself with this skill, which she had so far never had to put to use. We both agreed that it would be wise not to lose it. So, on the shore of Lake Salla, I dictated Balzac's *La cousine Bette* to her, with frequent pauses to take in the beautiful view of the surrounding hills. We would also stop our work to take a dip, or to eat at the hotel, where the food was simple and plentiful. Little did I guess that our friendly hosts found our diligence disturbing; seeing us dictating and typing copious notes while apparently scrutinizing the surrounding area, they came to the obvious conclusion: that I could only be a Soviet spy. From this time on, the central police in Helsinki kept a file on me.

The rest of the month of August was devoted to a trip which took us to the far north of Finland, to Petsamo (later annexed by the U.S.S.R and now known as Pechenga) on the Arctic Ocean. There too I did not fail to take a dip, in a cove with silky sand and water that was marvellously limpid but so icy that I could not stand it for long. The sun barely dipped below the horizon at midnight: we hardly knew when to sleep and when to wake. Reading the newspapers was out of the question; the only papers available at these latitudes were in Finnish, for Swedish is spoken and understood only in southern Finland. Still, I had the feeling I understood Finnish when in a hotel lobby I saw the headline: RIBBENTROP... MOLOTOV... MOSKVA. These words announcing the signing of the pact between Stalin and Hitler were blindingly clear.

We returned to Helsinki just before the declaration of war. When it was announced in special editions of the newspapers, Eveline and I were sitting in an outdoor café on the Esplanade, not far from the main theatre in the city. We felt as if we had lost a beloved friend. Eveline decided to stay for some time longer with me. Mail was still arriving from France, with some delay, and she continued to receive news from her family. Ahlfors, who lived in Munksnäs, a suburb of Helsinki, helped us find a nice furnished room with the use of a kitchen, not far from his house. We took walks in the parks along the Baltic Sea, feeding the tame little squirrels who would come sit on our hands. I wrote a number of letters to friends, in America and elsewhere, to keep them informed of my situation. To simplify matters, I portrayed myself as a conscientious objector.

I was beginning to have misgivings about the safety of my chosen place of refuge. Stalin was behaving more and more arrogantly with regard to the Finns. It has been claimed that the major concern of the Finnish general staff, in case of war, was how to feed the population of Leningrad once the Finns had occupied that city. Even if this is an exaggeration, it gives an accurate picture of the unfounded confidence the Finns had in themselves, for it is highly doubtful that they had any intention of calling on Hitler for help; they could hardly have been unaware of the price they would have had to pay for such a move. I was also told that they had deliberately decided to use Soviet-caliber artillery, so as to ensure continuing replenishment of their munitions from con-quered enemy stores.

In any event, Eveline could not remain any longer away from her son Alain, who had been entrusted to his grandmother's care. Communica-tions with France were liable to be cut off at any moment. She decided to return to France by the only route still open: by train through Sweden and Denmark, by plane from Copenhagen to Amsterdam, and again by train. We wondered whether we would ever see each other again; even this was uncertain, but we held on to some irrational hope for the future. We parted sadly on October 20, 1939. I considered moving to Sweden to await further developments – but it was already too late for that: because of my irregular standing with respect to the French authorities, I could not obtain the Swedish visa which had become necessary now because of the war.

November passed fairly uneventfully. Eveline had been reunited with her mother and her son in Parcé, the small village in the Sarthe department where her father had been born. Fearing that letters signed by me and arriving directly from Finland might compromise her, I had appealed to the kindness of Louis-Philippe Bouckaert, the Belgian physicist with whom I had become friends in Princeton, and who was now teaching at the University of Louvain. We agreed that I would send him unsigned letters meant for Eveline, and he would sign them before sending them on to her.

I generally dined with Ahlfors, who had sent his own family to safety in Sweden. In culinary matters I was, and still am, utterly illiterate. For my lunches, Eveline had left me with a little book of recipes within the grasp of my limited abilities. I continued my work for Bourbaki, though without any great feeling of conviction, and I provided myself with reading at the excellent municipal library. Having access, fortunately, to a good radio in my furnished room, I could not resist spending a fair amount of time listening to broadcasts from London and Berlin. I listened to the retransmission of Hitler's visit to Munich that had been interrupted by a bomb. This unsuccessful attempt on his life was not explained until some time later. I heard a report from Berlin on the Kaiser-Wilhelm Institut für Botanik. At the end of his visit, the journalist asked the professor, "So, Herr Professor, I see that you're going on with your work?" "Yes," answered the professor, *"hier geht es noch im guten alten Stil* (here everything is the same as in the good old days)." I hoped he would not suffer for having made this remark.

I also listened to good music. Once in a letter to Eveline (that is, to Bouckaert, to be forwarded to Eveline) I wrote: "...I had the good fortune to chance upon a Mozart quintet (a string quartet plus clarinet) in a marvelous performance rebroadcast from Berlin. Mozart's music, even at its most beautiful, often gives an impression of some being who, though very far above us in his incomprehensible serenity, nevertheless stops to remember us for a brief instant and comes within our reach, with gentle mockery and tender pity, to transcribe a fleeting message for us. But sometimes, in certain quartets and quintets, and in certain parts of *The Magic Flute,* this same being, without a thought for us, communicates with his fellow beings, and what we hear then is a world unknown to us, a world of which we are allowed only a furtive glimpse."

This was the moment of calm before the tornado. On November 30, the Russians dropped the first bombs on Helsinki. That day, in my suburb, I realized only that something out of the ordinary was taking place.

I followed my neighbors, who were heading for the countryside. Towards noon the air raid alert ended and I returned home. I approached a nearby square, unable to make out what was happening there. In fact, there were several anti-aircraft machine guns. My myopic squint and my obviously foreign clothing called attention to me; I was taken to the nearest police station, where the central police were contacted by telephone. There was already a file on me, from the Lake Salla days. I was immediately taken to the central police station and put in jail. My major preoccupation just then was with the nice slice of ham that I had waiting for me at home – an unaccustomed luxury that I had planned for lunch that day.

I must have spent four or five days in jail. The police conducted a search of my apartment, in my presence. The manuscripts they found appeared suspicious – like those of Sophus Lie, arrested on charges of spying in Paris, in 1870. They also found several rolls of stenotypewritten paper at the bottom of a closet. When I said these were the text of a Balzac novel, the explanation must have seemed far-fetched. There was also a letter in Russian, from Pontrjagin, I believe, in response to a letter I had written at the beginning of the summer regarding a possible visit to Leningrad; and a packet of calling cards belonging to Nicolas Bourbaki, member of the Royal Academy of Poldevia, and even some copies of his daughter Betti Bourbaki's wedding invitation, which I had composed and had printed in Cambridge several months earlier in collaboration with Chabauty and my wife. Along with the Salla file, all this amounted to an incriminating bundle of presumptions. I underwent a rather calm, if lengthy, interrogation at the police station. That day I saw just how deeply ingrained in me the calling of "professor" was. During the interrogation, which was conducted in German, the policeman was trying (rather feebly) to catch me contradicting myself. At one point he said, *"Sie haben gelügt"* ("you have told a lie" – but with a grammatical mistake). My immediate reaction was to answer with: *"Nicht* gelügt, *man sagt* gelogen (it's not *gelügt,* the correct form is *gelogen*)." Fortunately, he did not appear to be offended. For corroboration of what I told them, I referred the police to Ahlfors and Nevanlinna. The latter, as will be seen, was no longer in Helsinki; as for the former, when he inquired after me, he was told, rather brusquely, that he had best not get involved. In my situation, I could not appeal to the French legation; nevertheless I was taken there. The legation official whom I saw was the nephew of Paul Dupuy, who had long been the popular school secretary at the Ecole Normale. This man appeared extremely jittery; perhaps I wrong him, but he seemed to me to exhibit every symptom of being scared out of his wits: bombs had fallen not far from the legation. I had to confess my illegal status to him. He called me a spy and a deserter, though I later learned that according to French law I was in fact guilty of no more than failing to report for duty. In sum, he made it clear he was washing his hands of me entirely. I was later told that he went on to a successful diplomatic career under De Gaulle.

I cannot complain about the hospitality extended to me by the Finnish police. The cell, a small one, was clearly not intended for lengthy internments. I shared it with two others. One of them, a pirate by the looks of him, had sailed the China Sea. He spoke English fairly fluently, and taught me a pidgin English version of Heine's *Lorelei,* which began *"Me no savvy..."* I have always wished I could remember the rest.

The most aggravating circumstance was that a strict blackout had been imposed on the city, but as no preparations for this had been made in the cells, we had to spend the first day and perhaps the second without any lighting at all. Given the latitude and the season, this meant that we could do nothing but doze on our cots and eat the meal served to us in the near pitch dark. The second day the little window of the cell was painted over in dark blue, so we could light a lamp. Also, from that day on, instead of being given meals in the cell, I was taken to a nearby restaurant. I think it was on the third day that, on the way out, I saw police files being loaded onto trucks. I surmised that Helsinki was being evacuated because Russian troops were approaching, and thought it likely that I would be executed to spare my captors the bother of taking me with them. I was in a not unpleasant state of passive lucidity; this manner of death seemed to me a miniature version of the stupidity that had infected all Europe at the time. When I was led out of the jail the next morning, I did not even wonder what fate awaited me.

What had happened, according to the story told to me by Nevan-linna twenty years and one wife later, was as follows. He was, I believe, a reserve colonel on the general staff, and quite well-known in both governmental and military upper echelons. The day Finland entered the war, he left to take up his post. For obvious geographic reasons, he was stationed not far from Helsinki. On December 3 or 4, he was present at a state dinner also attended by the chief of police. When coffee was served the latter came to Nevanlinna saying: "Tomorrow we are executing a spy who claims to know you. Ordinarily I wouldn't have troubled you with such trivia, but since we're both here anyway, I'm glad to have the opportunity to consult you." "What is his name?" "André Weil." Upon hearing this, Nevanlinna told me at this point in his story, he was shocked. "I know him," he told the police chief. "Is it really necessary to execute him?" "Well, what do you want us to do with him?" "Couldn't you just escort him to the border and deport him?" "Well, there's an idea; I hadn't thought of it." Thus was my fate decided.

I was therefore escorted to the train station and shut up in a locked compartment with three other inmates. The trip was long, so we had plenty of time to get to know each other. When my neighbor, who spoke German tolerably well, introduced himself as a former Soviet general, I was somewhat surprised, even under these circumstances in which virtually nothing came as a surprise. As far as I could gather, he must have been a cavalry officer in the Tsar's army; during the civil war he had sided with the Soviets and, he said, had been commissioned General. Later I was told that during the civil war, high ranks, including that of General, were passed out rather freely. In any case, after the civil war he had come to Finland to set up a small business. His past rendered him suspect enough to be shipped out of Helsinki. We tried to guess at our destination. The most plausible conjecture was that we were being sent to a concentration camp for the duration of the war. Perhaps this is what happened to him. He told me all kinds of stories, of which the best (one he claimed to have witnessed firsthand) was about Trotsky, who was irked by Voroshilov's arrogant behavior toward him. Trotsky took advantage of a council of war over which he was presiding in Petrograd and, raising his voice, addressed Voroshilov thus: "Commander of the Tsaritsyn front! Comrade Voroshilov!," then, as if giving orders, "ATTEN-TION!" At this, according to the story, Voroshilov, frozen on the spot, stood to attention, and this marked the end of his insolence. *Se non è vero...*

My cellmate also taught me a Russian proverb. One of the seats in the compartment formed a commode emptying onto the tracks. With the utmost courtesy, he excused himself for having to disturb me by making use of it, and added: "As the Russian proverb says, a piss without a fart is like a wedding without music."

At one stop I was taken off the train to spend the night in a large prison. I took leave of my general; we wished each other luck. I have only the vaguest memories of the rest of the trip. I was again sent on by train, then made to get off. A policeman gave me back my passport and the wallet that I had had on me when I was arrested; he pointed to a large bridge, and motioned to me to walk in that direction. At the other end of the bridge was a Swedish gendarmerie post.

Arctic Intermezzo

I was at Haparanda, deep in the Gulf of Bothnia near the Arctic Circle. The gendarmes were not there exclusively for me: Haparanda was the principal way station for all those leaving Finland on account of the war and who were either unwilling or unable to travel by boat or plane. The Swedish had sent troops and gendarmes to patrol the border.

I had to tell my story once again. In Helsinki, I had revived and improved the basic Swedish I had learned when staying with Mittag-Leffler; by resorting to German and English when necessary, I had no trouble making myself understood. The officer interrogating me seemed to find my situation a simple one, until he asked me whether the Finnish police had mistreated me. I assured him that they had always behaved quite properly towards me. This answer, I saw, puzzled the Swedish officer. Many Swedes tend to regard the Finns as savages.

I was more or less penniless: my "war chest," along with all my baggage, was still in my room in Helsinki; I had only what I was wearing. Fortunately, these were my winter clothes, which I had brought with me the previous summer thinking my stay in Finland might be a lengthy one. At first the police paid for my room and board in a private home; after a few days they transferred me, not without apologies, to the local police station, which was outfitted with three large cells normally used exclusively as nighttime shelter for local drunks. I was free under surveillance, which meant I could go eat my meals in a well-patronized small restaurant ten minutes from where I slept, and I could even take a walk along a predetermined route. Seeing that I had no hat, a kindly gendarme gave me a wool cap. The temperature was no higher than $-20°$ Celsius, but because it was a dry cold, and the air was usually quite still, it was tolerable and even invigorating. In this season the sun was above the horizon for precious few hours of the day, but the prolonged twilight made the day several hours longer. Although the police station was rather lacking in conveniences, my stay there was not unpleasant. Once, when the gendarmes had to leave for some time, they asked me to answer the telephone for them.

I was not alone in my shelter for long. After a few days, I was joined in my cell by a Finn of approximately my age who spoke excellent English. A little later, an older man was placed in the next cell. He was a Finnish Communist deputy, a member of Kuusinen's puppet government, which was supposedly to have taken power in Finland if the beginning of the war had touched off a general revolt in support of the communists: this, apparently, was the outcome that the Finnish communist party had predicted to Stalin. If he believed this, it showed that he was as poorly informed about Finland as he had been about Germany at the time Hitler came to power. There was still one cell left for the drunks, whose occasional presence was no bother to us. Our communist spoke only Finnish, and even in Finnish he was rather taciturn. The only one left for me to talk to was my cellmate, who turned out to be pleasant and friendly. When the war broke out, he had been the head of the currency exchange section at the Ministry of Finance in Helsinki. Mobilized as an officer when war was declared, he had not been physically able to stomach the sight of the massacre to which the Russians had sent their troops in the early days of fighting: by all accounts, it was a horrendous scene of slaughter. There is reason to think that the Russians had sent their worst troops to Finland – whether because they had counted on support from a popular Finnish uprising; or because they had greatly underestimated the valor or the patriotism of the Finnish army; or again, as some have claimed, because of some Machiavellian calculation on Stalin's part.

At the suggestion of my cellmate, I wrote letters both to the French legation and to the Finnish authorities in Helsinki, in an attempt to recover my belongings. All that came of this request was that my trunk was eventually delivered to the French legation, whence it was forwarded to me in Brazil in 1946. There I was indeed happy to find in it my 1528 edition of Sophocles, the typographical masterpiece of Simon de Colines, which was very dear to me and which I had long since given up for lost. By the same channels as before, I transmitted news to my family in Paris. I also corresponded with some mathematicians in the hope of finding some way out of my situation. Viggo Brun, in Norway, hastened to send me money, which I was delighted to be able to repay after the war. Though he had once been a militant pacifist, now he did not hide the fact that he disagreed with the principles that had inspired my conduct; nevertheless he assured me of his wish to help me. For him too, Hitler represented absolute evil, and the fight against Hitler took precedence before all else. He took great pains to obtain a visa for me to come to Norway; it was a stroke of good luck that he did not succeed, for there I would inevitably have fallen into German hands a few months later.

In December, the police received orders from Stockholm to the effect that I could not stay in Sweden. Either I was to have the French Embassy send me back to France, or I was to be returned to Finland. I had neither the money nor the visa to go to any country whatsoever, but I had to choose the lesser of two evils. I contacted the French embassy, which no doubt consulted Paris on what to do about me. In January, I received a train ticket for Stockholm. I enjoyed several days of temporary freedom, during which time I met the mathematician Cramér. He too showed understanding and generosity, taking me out to eat in one of the city's better restaurants. I went back to Skansen, the lovely zoo and open-air ethnographic museum that Eveline and I had visited with such delight just months earlier. I conjured up memories of her and of Nils Holgersson who had enchanted me as a child. At the embassy, a legation officer was friendly and sought to reassure me as to the fate that was awaiting me. In exchange for my passport he gave me an appropriate travel document, along with a train ticket for Bergen and a ticket for the ship from Bergen to Newcastle.

Under Lock and Key

At the end of January, then, I sailed from Bergen, in an old bucket of a six-hundred-ton ship. A pall of disaster hung over the crossing: there were radio announcements of boats sunk before and behind us. No doubt they had struck floating mines. There were instructions to wear life-belts day and night. I was perhaps the only one on board never to don one – not out of some foolish sense of bravado, but because the chance of surviving a shipwreck seemed virtually nil to me: the North Sea is not known for its clemency in January. In fact, the crossing was so rough that I spent 36 hours dozing, without a bite to eat. As we neared Scotland, the weather grew calmer. Once there we had to wait three days for the Admiralty's orders to dock, and we witnessed the unwonted sight of a throng of boats at anchor, all immobilized by the state of war.

I do not recall exactly how I expected the rest of my journey to transpire: perhaps I thought that, thanks to the generosity of Viggo Brun and Cramér, I had enough money to get to France on my own and turn myself in to the authorities there. I do remember that I was hoping to visit my friends in Cambridge on the way. All such illusions were dispelled as soon as the boat had docked: the police were waiting for me. I had not yet suspected that the French and British police were already working hand in glove. Precisely how had the British police been notified of my presence on this ship? I did not give much thought to this question at the time, nor did I ever find out.

Taken into custody even before disembarking, I was interrogated for quite some time by a young inspector (if that was indeed his rank) with an Oxonian air about him. He examined and confiscated the various papers I had in my possession. One of them mentioned that Bertrand Russell had taken an interest in my case. The inspector asked me, "Is that Lord Russell?" Indeed it was: he had inherited the title some years previously. The interrogation turned into a discussion on the pros and cons of conscientious objection. After an interruption, my questioner returned to me saying, "What can we argue about now?" In sum I was taken to the train, my wrist tied with a leather thong – "like a dog," smiled the policeman to whom I was attached. Thus did I make the journey from Newcastle to London on the night train. When I had arrived and been given breakfast at the train station, I was taken to my new quarters, at a local police station, for yet another interrogation. Here regulations prescribed an hour's walk daily for every prisoner. So the bobby on duty took me through the streets of London and, at my suggestion, to the quays along the Thames. He seemed well pleased to be assigned as my escort. His all-time pride and joy had been to escort Léon Blum when he had come to London on an official visit as Prime

Minister. The policeman also had anecdotes of another variety, such as the night patrol when he had approached a car with all its lights out parked on a dark side street. He was not surprised when his flashlight revealed a couple reveling in an advanced state of undress; but he was slightly taken aback when the man shouted, "Get away, you're embarrassing my lady-friend."

From London, I was taken – still under escort, but this time not tied at the wrist – by train to Southampton, where I was introduced to a compatriot. Though he seemed to have difficulty communicating with his English colleagues, he ignored my offer to interpret for him. As soon as we were alone he explained, "I understand English perfectly; but in my line of work, it's best to act as if you don't know the languages you know best." He was an agent with the French intelligence service. While leading me on board the ship that was to take us to Le Havre, he queried me about my situation, and then asked, "But with all these difficult problems, haven't you contemplated suicide?" I told him that the idea had never so much as occurred to me. "Well," he said, "it's just that I don't want any trouble. If you do have any such notions, I'm going to clap you in irons straightaway." I assured him that this would be unnecessary. By now we were in a comfortable cabin with two berths. He said, "I have work to do on board. I'm going to lock you in here. I'll be back tomorrow morning."

As promised, he released me in the morning and took me to the first-class dining room, where breakfast was being served. At the next table were Paul Langevin, Maurice Fréchet, and the physicist Sadron, my colleague and friend from Strasbourg. I asked my companion, "Would you like me to introduce you to a member of the Academy?" He seemed flattered by the question.

When I was at the Ecole Normale, my classmates and I respected Langevin for a number of reasons, among which was the fact that he had never stooped to be a candidate for membership in the Academy of Sciences. Some time before the war, he had resigned himself to doing so, supposedly at the behest of the Communist Party: then the communists could publicize meetings "attended by Paul Langevin, member of the Academy." Once it even happened that a party poster announced "Paul Langevin, Member of the Academy, Nobel Prize Laureate." Langevin probably deserved the Nobel Prize as much as anyone, but the fact is that it was never awarded him. To the great chagrin of his friend Paul Rivet, this eminently worthy man let himself be used without a word of protest.

I therefore introduced my escort to Langevin. We were supposed to disembark directly after breakfast, but because of a thick fog we remained on board for the better part of the day. Fréchet took me aside to tell me the following: "In London, people were saying you had been caught in the act of being a spy in Finland. But I didn't believe it. If that had been the case, the Finns would have shot you. They didn't, so you couldn't have been one." His axiomatic reasoning was impeccable.

Sadron explained to me that he and his colleagues were returning from a scientific mission in London. We lamented the dreadful times. I had another long conversation with my escort. "You Jews are really amazing. Once I met a Jewish lawyer in the army, during the war of 1914-18. Our regiment was marching through enemy territory. One day he said to me: ,What am I doing here? I am a free man. If I feel like leaving, I can leave.' And he took off. He came back the next day. I asked him, ,What have you been doing since yesterday?' He answered, ,A lot of thinking.'"

Before leaving me, the agent concluded our conversation thus: "In Le Havre, I am going to turn you over to the gendarmes. They are not as broad-minded as I am. You had better notify your family that you've arrived. Write the letter and give it to me; I'll post it myself." So I wrote a brief letter to Eveline and another to my parents, saying that the most urgent problem was to find me a lawyer. I asked Sadron to transmit the same message verbally, and I added: "Make sure the lawyer isn't named Levy or Cohen." I do not know whether the message was transmitted, but some time later I received a telegram informing me that my "defender" was named Maître Edmond Bloch.[1]

When we had docked, the gendarmes took me to the jail in Le Havre. In comparison with the facilities I had visited in Finland, Sweden, and England, this French jail seemed fairly dirty and poorly kept up. Thoroughly frisked and stripped of what little remained in my pockets, I was put in a brightly lit and roomy cell, designed for several inmates but empty except for me. My only diversion was deciphering the inscriptions and the awkwardly obscene drawings decorating the whitewashed walls. The next day, when I was given my lawyer's name, I wrote to him, stating that my morale would not withstand solitary confinement for very long. I figured that, since my letter inevitably had to be read by the prison administration, some improvement in my treatment might result, and that if need be my lawyer and my family would use the letter to that end. Later I found out that my parents and my sister had taken my complaint seriously and were quite worried. In any case, it had the desired effect upon my captors: I was moved to the neighboring cell, which already held two prisoners. One was a poacher by trade, who recited stories of village sorcery that he was prepared to swear were true. At his trial, he had told the examining magistrate: "Your honor, when it's not hunting season and

---

1 *Translator's note:* The term *"Maître"* is the honorific commonly used for lawyers in France.

you eat pheasant, where do you think it comes from? It comes from me." This remark had not saved him from being sentenced to six months, which he was now serving in Le Havre. He could have easily reconciled himself to his fate, had it not been for the lack of female companionship. But he had taken pains to remedy the situation the instant he was released from prison: his wife would be coming to wait for him, and a room had already been reserved in the hotel across the street.

According to regulations, prisoners were supposed to be allowed fresh-air walks and access to the library. At Le Havre, there was no possibility of walks outdoors, and there was no library. After repeated requests on my part, someone managed to dig up, God knows where, half a dozen books in English which had been left behind by some earlier inmate. Thus it was that I read Willa Cather's wonderful novel *Death Comes to the Archbishop,* as well as a fascinating book that I have never since come across: the memoirs of a Czech locksmith who, smitten with wanderlust, had traveled through Russia and then Siberia to the far north. There he had served as locksmith, engineer, dentist, and even elected judge for an immense territory inhabited primarily by Eskimos, with very few whites. As judge, his duties were to enforce the sole law, which prescribed capital punishment for selling alcohol to the Eskimos, and to keep a record of the executions. He described the approach of spring, which was heralded just before the end of the polar night by flocks of birds migrating northward: high in flight, the birds were already illuminated by the sun, while it was still night on the ground.

The warden in charge of the prison was known for his brutality. I was not very worried on my own account. For one thing, I had been registered as "Lieutenant Weil," and my rank afforded some protection from the prison guards. Nevertheless, I was glad when told I would be transferred to Rouen, where, in the company of two gendarmes, another inmate and I were taken by train. When a gendarme approached the other prisoner with handcuffs, he instinctively recoiled. "Come on," the gendarme told him, "there's no shame in it."

Thus it was that in mid-February, I began my stay in the military prison in Rouen, ironically known as "Bonne-Nouvelle"[1] after the neighborhood where it was located. I spent nearly three months there. This prison was quite crowded, primarily with men arrested for "defeatist talk" on the order of "Better a live Kraut than a dead Frenchman." The Daladier regime, the phoney war, the obvious inanity of the official slogans ("We will win because we are the stronger") had unleashed a tide of demoralization that military justice was attempting in vain to stem.

---

1   *Translator's note:* "Good News," as in the Gospel.

For the first few days I had a cellmate, but as soon as Eveline, my parents, and my sister were visiting regularly and bringing me books, I managed to convince the liberal-minded prison director to put me in an individual cell where I was allowed to keep not only the books that my family would bring me each week, but a pen, ink, and paper. The cell, very long and narrow, was lit by a high window which showed a patch of sky through the bars. The furnishings included a bed, a chair, a small writing table attached to the wall with a shelf over it, a sink, and a flush toilet. The door, like that of a safe but pierced with the usual peek-hole, opened with a loud rattling of keys and a grinding of hinges – aptly illustrating the expression "as graceful as a prison gate." I had with me the *Bhagavad Gita*, the *Chandogya-Upanishad*, novels by Balzac, and, on my sister's enthusiastic recommendation, Retz's memoirs. I also had what I needed to resume my mathematical work. Soon I was receiving a large amount of mail, which had to be read first by prison officials and by the examining magistrate; sometimes there were censored passages. Almost everyone whom I considered to be my friend wrote me at this time. If certain people failed me then, I was not displeased to discover the true value of their friendship. At the beginning of my time in Rouen, the letters were mostly variations on the following theme: "I know you well enough to have faith that you will endure this ordeal with dignity..." (sometimes preceded by the theme, "You know I do not agree with your views, but...") But before long the tone changed. Two months later, Cartan was writing, "We're not all lucky enough to sit and work undisturbed like you..."

I learned to write longhand again. I went back to the Bourbaki report on integration, which I had left off in Finland. Freymann, showing himself a true friend, was eager to make up for the unexplained delay in publishing my manuscript on integration in groups, which I had given him in 1937; he sent me the page proofs to correct in Rouen. This work was dedicated to Elie Cartan, and it was also in Rouen that I wrote the dedication. One day my parents, who had come to Rouen for their weekly visit, stopped on their way in to greet the prison director, as was their wont. He told them, "Your son is doing well. He finally finished an introduction which cost him a great effort, but now he's happy with it." The director knew all about my work from reading my mail.

In keeping with the regulations, I had my half-hour's walk each day. It was not exactly like the prisoners' walk portrayed in Van Gogh's painting. Each prisoner was assigned a sector of a large circle, in the center of which the guard tower was located. By walking fast one could get a bit of exercise. In my cell I would also do some exercises every day. There were a few trees growing on the walk. When their leaves started to come out in spring, I often recited to myself the lines of the *Gita: "Patram puspam phalam toyam...*" ("A leaf, a flower, a fruit, water, for a pure heart everything can be an offering"). Only once, by accident, I happened to share my walking area with a fellow prisoner, who was miserable from the deadly boredom of internment. He was not aware that while awaiting trial he had the right to have books brought to him; he told me he loved reading. If every detainee were informed of his rights, much pointless suffering would be avoided. Life is a composite of trivial details: perhaps this becomes clearer in prison than anywhere else. The regulations in Western nations are generally far from inhuman; but to those whose lives are thus determined, whether these regulations are enforced with cruelty or kindness, or simply with a closed or an open mind, can make all the difference in the world – the difference between misery and a tolerable existence.

Most certainly, I never experienced any but the most benign aspects of incarceration. The wardens, for the most part seasoned professionals, were by no means sadistic; no doubt their motto was "Avoid trouble." Experience had taught them enough basic notions of practical psychology to be able to recognize which of their wards could be trusted and which could not. With me, for example, the weekly cell search, for razor-blades, files, or other tools of escape, was merely a formality. Once, however, it transpired otherwise. The door opened, with its inevitable massive noise. The regulations prescribe that when the cell is opened the prisoner is to rush to the far end of the cell and stand to attention. I therefore "rushed," rather lackadaisically I suppose, and struck a posture vaguely resembling "Attention." The guard on duty, whom I knew, was a small, somewhat choleric man. He began to feel every inch of the mattress, then underneath the table. No doubt he saw an ironic smile form on my lips. He drew himself up to his full height, saying, "I learned my trade from old prison guards. I am performing my job as they taught me to." This man was no mere prison guard; he was a guardian of tradition.

In the meantime, the investigation of my "case" continued. As may be inferred, it was actually quite simple. The investigating officer was named Le Lem. He was, I believe, a civilian magistrate dressed in a military uniform for the duration. It appeared to me that what he was called upon to do in the line of duty – which consisted primarily of prosecuting "defeatist talk" – seemed stupid to him. One day when he had summoned Eveline and she spoke of our stay on a Finnish island, he said to her in a voice tinged with nostalgia: "And what did you do on that island? Did you go fishing?"

I also had several visits from Maître Bloch, my defense attorney. After our first talk, there was really nothing left to say about my case; we chatted, and he kept me up to date on world events. I was deeply saddened when he told me how the war had ended in Finland. Yet it was not from

Rouen, 29/II/40

Ma chère sœur,

J'ai bien reçu tes lettres, celle qui était allée au Havre et celle que tu m'as envoyée ici. Je suis fâché que vous ne puissiez me voir pour l'instant ~~████████████████████~~ ~~████████████████████~~. Mais je pense que cela ne tardera pas.

Je suis attelé, pour l'instant, à la correction de mes épreuves: cette besogne minutieuse et mécanique, que je ~~n'interromps~~ ne romps guère que par la lecture de Balzac, me convient assez bien ces-jours-ci, après tant de temps passé sans aucune occupation. Je pense avoir terminé bientôt, bien que le fait de ne pas avoir de lumière le soir, et de devoir par conséquent se coucher avec le soleil, retarde mon travail. J'écris aussi, pour le même fascicule, une épître dédicatoire à papa Cartan, à l'ancienne mode, ce qui permet, sous un langage un peu pompeux et solennel, de faire passer ce que je veux dire de lui. Je t'en communiquerai le texte, pour retouches éventuelles, ainsi qu'à H. Cartan.

Quant à parler à des non-spécialistes de mes recherches ou de toute autre recherche mathématique, autant vaudrait, il me semble, expliquer une symphonie à un sourd. Cela peut se faire: on emploie des images, on parle de thèmes qui se poursuivent qui s'entrelacent, qui se marient ou qui divorcent; d'harmonies tristes ou de dissonances triomphantes: mais qu'a-t-on fait

Letter from A.W. to Simone Weil, February 29, 1940

him that I learned of the Narvik expedition: I heard this from a fellow inmate who whispered in my ear on the way to the showers: *"Ça chie en Norvège."*[1]

In April, I made what I had reason to believe was significant progress on correspondences between algebraic curves. There were still quite a number of gaps to fill in; in particular, I was still seeking the proof of a certain elusive lemma. My sister maintained that my failure to come up with this proof was the result of Captain Le Lem's curse, and that I was doomed never to find it. In normal times I would have waited before publishing anything. This time, the future seemed so shaky to me that I thought it appropriate to publish a brief sketch of my results in the *Comptes-Rendus;* it was Elie Cartan who submitted the note.

All this is recorded in the letters I sent to Eveline between her visits. Most of the time she was in Parcé near Solesmes, with her mother and her son Alain, who was by then eight years old. Here is the greater part of these letters:

(March 4) [...] What can I say about myself? I am like the snail, I have withdrawn inside my shell; almost nothing can get through it, in either direction. I suppose this isn't very courageous of me, but the most important thing is to preserve oneself intact. Don't worry: inside this shell I am still the same, still yours; I will come out some day, and we will find each other again. Try to stay happy and brave [...]

(March 30) [...] I would really like to have had a description of the Holy Saturday ceremony in Solesmes – what is this about sacrificing a lamb? I don't understand it at all, and I would like details. I wish I could see the first buds in the countryside. Here, on the walk, if I crane my neck, I can make out the upper branches of some trees, and it's true, they are beginning to bud, but, well... I guess it will have to wait till another year. [...]

Actually, since I saw you, my arithmetico-algebraic research has got off to a good start. I have found some interesting things – to the point where I'm hoping to have some more time here to finish in peace and quiet what I've started. I'm beginning to think that nothing is more conducive to the abstract sciences than prison. My Hindu friend Vij. often used to say that if he spent six months or a year in prison he would most certainly be able to prove the Riemann hypothesis. This may have been true, but he never got the chance. [...] All this, along with my readings, keeps me plenty busy. I have run out of Balzac, but I reread bits from time to time; as I read them I often think of how much fun we would have had if I had dictated them to you; and of everything that would have made us laugh together.

---

1 *Translator's note:* a vulgar slang equivalent to "They are fighting in Norway".

These too are pleasures that will have to be postponed for a while. I am still reading Retz's memoirs,[1] but when I leave off for a day or two (something that occurs frequently) I have trouble picking it up again: I am always losing the thread of his triply- and quadruply-knotted intrigues and I get completely lost. One feels much sympathy for him, but nothing great is ever accomplished by such a convoluted mind. My sister was horrified when I asked her if he wasn't Neapolitain. It appears that the Gondi family came from Florence. But he doesn't have the grand simplicity of the Florentine mind. I am still convinced that he must have had southern blood in his veins (southern Italian, of course).

And then there are always my Sanskrit books. I am reading the *Gita,* in small doses as one ought to read this book. The more detail one absorbs, the more one admires it. I am curious to know what kind of impression it would make on a good Catholic. [...]

And then there is my correspondence with my sister; nothing could be easier than this at the moment, since it concerns only the most abstract subjects. Greek mathematics, for example: she has some ideas about that. The other day, she came to see me with my mother in the visiting room, and I was severely reprimanded by the guard on duty because I said three words in Greek. Of course we are not allowed to speak in any language but French. Having reflected on my work in a general way, I also wrote a very long-winded disquisition in the form of a letter addressed to her, since she has been hounding me forever to do that – but I warned her that she could not even try to understand without succumbing to superficiality. I fear she will take no heed of my warning. [...]

(April 7) My mathematics work is proceeding beyond my wildest hopes, and I am even a bit worried – if it's only in prison that I work so well, will I have to arrange to spend two or three months locked up every year? In the meantime, I am contemplating writing a report to the proper authorities, as follows: "To the Director of Scientific Research: Having recently been in a position to discover through personal experience the considerable advantages afforded to pure and distinterested research by a stay in the establishments of the Penitentiary System, I take the liberty of, etc. etc." [...]

As for my work, it is going so well that today I am sending Papa Cartan a note for the *Comptes-Rendus.* I have never written, perhaps never even seen, a note in the *Comptes-Rendus* in which so many results are

---

1   *Translator's note:* Jean-François Paul de Gondi, Cardinal de Retz (1613-1679), was one of the leaders of the Fronde, the aristocratic rebellion against the government of Ann of Austria, who was Regent to her son Louis XIV. The Cardinal's memoirs remain a classic of seventeenth-century French literature.

compressed into such a small space. I am very pleased with it, and especially because of where it was written (it must be a first in the history of mathematics) and because it is a fine way of letting all my mathematical friends around the world know that I exist. And I am thrilled by the beauty of my theorems, but of course that is hard to communicate to you. But I was thinking that when you go back to Paris you will see the 14-page letter I wrote to my sister; don't read the mathematical stuff, of course (I feel terribly guilty because my poor sister got dreadful headaches trying to read it), but there are a few comparisons that might amuse you. [...]

Mathematics notwithstanding, I would not mind alternating my fishing for theorems with fishing in the Sarthe River (under your supervision of course, since I know nothing about it), or else with some bicycle rides. Never mind, we'll get a chance some day. Here are some lines from the *Gita* that I like very much: "A leaf, a flower, a fruit, some water, whoever dedicates it with love, this love offering I accept with the devotion of his soul."[1]

The god Krishna is speaking. In Tamil it is written of him, "The bread we eat, the water we drink, the betel-nut we chew, all this is our Krishna." It is almost impossible to translate all of this – in everything to do with gods our language is slanted toward the idea of a personal god, that is totally foreign to the Indians.

I also amused myself by translating a chapter of my other Sanskrit book; it's just what Faust says in his famous opening monologue:

"Teach me, Venerable One!": with these words Narada came to Sanatkumara. He answered him: "Come to me with what you know; then I will teach you further." He told him: "Venerable One, I know the Rig-Veda, the Yajur-Veda, the Sama-Veda, the Atharva-Veda which is the fourth, the Legendary Lore which is the fifth, the Veda of the Vedas, I know ritual, mathematics, logic, morals, theology, brahmanology, demonology, astrology, herpetology. This I know, Venerable One. Such am I, Venerable One, knowing the science of books, but not the science of being. I have heard it said by such as you, Venerable One, that he who knows the science of being crosses over sorrow. I, Venerable One, am sorrowing. I ask you, Venerable One, teach me to cross over to the other side of sorrow." He told him: "All that you have named are but words. They are only words, the Rig-Veda, the Yajur-Veda, the Sama-Veda, the Atharva-Veda which is the fourth..." (and the whole list is enumerated all over again).[2]

After which he teaches the science of being. It is not quite up to our expectations; but that is because our minds have been spoiled. When this

1 *Bhagavad Gita,* IX: 26.
2 From the *Chandogya-Upanishad.,* Seventh Prapathaka, First Khanda. Translation modified from Robert Ernest Hume (Delhi: Oxford, 1983) 250–251.

text was written (it is very old – from around the 6th century B.C., I think) it had everything necessary for meditation. The Indian philosophers, who studied and commented upon these texts *ad infinitum,* also found everything they wanted in them – but that, I think, is sometimes because they put it there... Even nowadays, in many of the higher castes, a Hindu belongs to a certain philosophical school by birth; this does not stop him from believing whatever he wants, but if he is a good Hindu he will always claim that he is only interpreting the principles of his school, and that there is no contradiction.

If I get started on this topic, I won't finish for a year, and you may not find it terribly interesting – but I can hardly amuse you by describing the walls of my cell, which are the only landscape before my eyes right now; and of everything in the *Gita,* all I have to offer Krishna is water, and now and then a fruit – an orange or banana that they give me for dessert; sometimes, these last few days, a young leaf, all crinkled up still, that the wind has blown onto the walk – but no flowers (but then it's not yet the season; maybe you don't have any either). I would gladly offer a mango, but they are difficult to find here. [...] P.S. My father is afraid I will get writer's cramp.

(April 16) My Eveline, a few days ago I received your letter of April 6, full of flowers which were sorely needed in my cell (why did I imagine that you didn't have any yet?) and also of the garden scents of grasses, even snails. All the jonquil petals are lovely to contemplate when I pause from my mathematics – I do still pause from time to time. But you are becoming too knowledgeable for me: I haven't the least notion of what it is to stake pea plants; you will have to teach me, in theory for the time being and perhaps some day in practice. My sister says that when I leave here I should become a monk, since this regime is so conducive to my work. I will join the Benedictines at Solesmes, I will study Gregorian chant and I will sneak away to come visit you in your garden. But what will the good folk of Parcé say when they see a monk stowed away in your yard so often? Perhaps they will report me to the Father Superior?

There are no flowers here; the trees I see in the morning grow greener by the day, and on my walk this morning the air had a light springtime feel; but alas, it doesn't really reach as far as the cells, and I have to be satisfied with a very small ray of sunshine toward the end of the day. Things are a bit calmer on the mathematics front now; it's time to work out the fine points of the proofs, and this is never much fun. I have nothing new to read. Having finished Retz's memoirs a while ago, I am now rereading them, and they improve on the second reading: now that I am familiar with the events (which are rather clumsily narrated) I no longer have to rack my brains to follow the thread of these machinations, and I can pay more attention to his descriptive passages and to his general reflections, which are interesting. What do you think of Mademoiselle de Chevreuse, who treated her lovers like her skirts – "she would take them to bed when she felt like it and two days later she would burn them out of sheer spite." She is the young woman Madame de Montbazon is talking about when she tells Retz "that she could not understand how he could be amused both by an old woman, who was more wicked than the devil, and by a young one who was even stupider, if possible" (the old one was Madame de Guéméné). The conversation continues thus: "I was accustomed to her words, but, as I was not used to her sweetness, I found it interesting, although it aroused my suspicions, given the context. She was quite good-looking; I was not inclined to let such an opportunity escape me; I relented a good deal; my eyes were not torn out; I suggested that we go into her private chambers, but it was suggested that before all else we go to Péronne; thus ended our love affair."

And I continue, with as much enjoyment as ever, my reading of the *Gita*. I had often reread certain passages, but I had read it straight through from beginning to end only once, in 1923 or 1924. I am pleased with myself for remembering my Sanskrit well enough to read such a text – with constant assistance from the translation, naturally. The one attributed to Sylvain Lévi is, upon examination, rather uneven: there are portions which are entirely worthy of him (and that says it all), and others where the translation does not even seem accurate. Since this work, rather than being his own, was done by a student under his supervision and with his participation, perhaps he put his hand to it only in the parts he liked best, or when his student came up with the worst howlers. Besides, he always said that whenever only one translation is given for a line of Sanskrit, there is always at least one misinterpretation.

We've taken up our Bourbaki correspondence again, with a view to preparing a few more installments for publication. Having also recommended to Dieudonné that we start up our Bulletin again, I have just received some sheets entitled "The Tribe: An Ecumenical, Aperiodical, and Bourbacchic Bulletin," that opens with a letter beginning "To all of our brothers in Bourbaki, greetings and blessings" and ends as follows: "Let brains be de-rusted! Let pens scratch upon paper! Let the clatter of typewriters and the hum of presses carry the name of Bourbaki all over the earth! Amen." The rest of the letter is in the same style, only better.

As you can see, I have to call Retz, Dieudonné and *tutti quanti* to the aid of my faltering imagination; I had better stop for today. These few weeks of intense work have left me rather dazed. In other times this would have been the moment to take a little vacation – to go look at tulip fields, or frescoes, or just cherry trees in bloom. Now I have none of these – except the flowers in your letters (and these are far from nothing). The story of the Easter lamb and the Benedictine monks and nuns was interesting as well. I had never heard of these nuns. Do they also sing Gregorian chants? (and are there among them, for the monks, "young nuns of fifteen or sixteen"?)[1] [...]

P.S. Don't worry too much about explaining fractions to Alain. For some time elementary education has placed entirely too much emphasis on them; they really aren't so very important. The best that can be done, I suppose, is to learn to manipulate them mechanically, without trying to understand – especially for multiplication and division. On the other hand, since you speak of grammatical analysis, make sure he does this absolutely perfectly, because it is essential for learning Latin (which he would do well to start without delay). It is essential to be able to recognize the subject, direct or indirect object, etc., even in sentences with complicated word order. It is very important for what comes later not to let the least mistake slide by in this area. ("Logical" analysis, as it is called, is also important, but I suppose he doesn't do any of this yet.)

(April 22) [...] My mathematical fevers have abated; my conscience tells me that, before I can go any further, it is incumbent upon me to work out the details of my proofs, something I find so deadly dull that, even though I spend several hours on it every day, I am hardly getting anywhere. I have just finished rereading Retz; I have also finished my reading of the Gita, which is truly one of those texts that could be reread indefinitely. I finished all my Balzac a long time ago and returned the books to my family. Bourbaki is idle. I finished writing my observations on the continuation of our topology to Cartan and Dieudonné, and I told the latter that I would not continue to reply to the papers on integration with which he has been inundating me. Last week I rewrote my article for the *Revue Rose* and sent it off, completely reworked and three times as long as it was before; I have asked my sister to correct the page proofs and I also asked her, some time ago, to speak to Freymann to settle the unresolved questions with regard to my book, which I didn't want to discuss any further. Here I am, one might say, more or less on vacation. This place is not as well suited to vacation as it is to work; in this balmy weather, I would much rather be sitting on that bench surrounded by ivy, near the yellow flowers smelling of honey, where I would speak to you of Krishna – and of many other things, I'm sure. I am not surprised that instead of reading, you are busy following the trees and plants in their springtime changes. I am very glad that you have spent this season in the country, and in a region where spring

---

1   *Translator's note:* The words *"une jeune nonne de quinze à seize ans"* occur in a French folk-song.

is really spring – think of Strasbourg, where it was summer right away, despite the flowers. "Of all the seasons," says Krishna, "I am the season of flowers." But he doesn't tell us which flowers, and in India there are huge differences between forest flowers and mountain flowers (in Kashmir I once camped in a valley called Sonamarg, "golden alp," so thickly did it bloom with yellow flowers in April), or again the flowers of the plains – these are few, but there is a big tree with magnificent blossoms, whose name I have forgotten. These large blooms appear long before the leaves, and the tree looks as if it has caught fire in the sun, which is already quite hot in spring and soon becomes scorching. There are also little green parrokeets which are like flowers that move and even chatter – but they have no season. And there are garden flowers, but these require so many days of labor by the water-carrier and the gardener... Have I ever told you that in Bombay people who have no garden often have a gardener, whose job is to keep his master's apartment in bloom by cultivating good relations with gardeners who do have gardens? And flowers, lots of flowers, are also necessary at the arrival and departure of important guests, whose necks are draped with perfumed garlands: ordinarily only two or three garlands, but for special occasions, as many as the victim can carry and still stand up. I have seen people stooping under the weight of these garlands. [...]

Serving the Colors

I was "tried" on Friday, May 3, 1940. That morning, the house barber came to spruce me up. He was one of the inmates, and had come to shave me once or twice a week since my arrival in Rouen: naturally, being myself an inmate, I was not allowed to use a razor. This man had taught me a barber's saying, "well lathered is half shaved," which I have oft repeated to myself since then. This barber was an Italian who owned a brothel in Rouen. His madam had turned him in, he said, in the hopes that he would be deported after his jail term and that she would become sole proprietor of the establishment.

The trial was a rather badly-acted comedy. The judges, all of them officers, were in uniform, as was the military prosecutor. Maître Bloch cut a majestic figure in his black robes, against which the red ribbon of the Legion of Honor stood out. My family was present. With his characteristic kindness, Elie Cartan had agreed to come testify on my behalf. Technically, I was accused of failing to report for duty, rather than of desertion, since I was not actually in the military when I committed my offense. Maître Bloch's robes with their dramatic sleeves gave him an edge over the prosecutor. Apart from this detail, the indictment and the speech for the defense seemed to me equally uninspired. No one paid any attention to Elie

Cartan, who was not even asked to sit down. Before pronouncing the sentence (the "clubbing," in prisoners' slang), the presiding judge, in keeping with the prescribed ritual, asked me if I had anything to say. This was no time for a show of bravado. I said that if I had to don a uniform, I was ready to do whatever the army asked of me; but I could not bring myself to speak of repentance.

Besides, anything I might have said, and indeed whatever Maître Bloch or anyone else might have said, would not have influenced the outcome any. The following winter, my friend Cavaillès was to tell me that at the cipher office of the Ministry of War in Paris, he himself had been the one to code the telegram that prescribed my sentence to the tribunal. There is often cause to doubt the autonomy of "justice" in general and "military justice" in particular, but it is not often that one has the opportunity to catch it red-handed, as it were.

I was therefore sentenced to the maximum term for the offense, that is, five years in prison, and was of course demoted from my officer's rank. Immediately after the hearing, the prosecutor met with the day's convicts (there were three or four of us) to inform us about our right to appeal. Then, seemingly addressing himself to me in particular, he indicated that we could also petition for a suspended sentence, in exchange for serving in a combat unit. He seemed to suggest that such a request would meet with favor in my case. Maître Bloch suggested appealing my conviction, an idea that struck me as stupid. I took the 24 hours that I was allowed to think it over and submitted my request to be sent to a combat unit. Little did I know how wise this decision was – if it is really true that one month later, when the Rouen prison was evacuated as the Germans advanced, the wardens shot the prisoners there, rather than risk slowing their flight with this extra burden. Such, at least, was the rumor that circulated.

Returning to the Bonne-Nouvelle prison, I had to exchange my own clothing, which I had kept until then, for prison garb. The guard overseeing this operation said to me, "They make people into scientists, and then look what they do to them. Poor France!" Except for the clothing, my daily routine continued unaltered. The following Saturday, I was informed that my request for a suspended sentence with service had been granted, and that as of that very moment I was free – and summoned to Cherbourg for induction on Monday.

Eveline had planned to come visit me in prison the next day, and I knew that she would be arriving in Rouen that evening, to spend the night in a pension there. I managed to find the place, filled the room with flowers, and surprised her there in the evening. Since October, we had seen each other only through the double row of bars in the visiting room at Bonne-Nouvelle. The next day my parents and my sister, whom we had contacted by telephone, came to join us, and we spent the day walking along the Seine in the lush green Norman countryside: the apple trees were in bloom. The German advance had just been launched – it was probably no coincidence that I had been so quickly metamorphosed into a soldier. None of this prevented my mother from describing the improvements she was planning for the apartment on the Rue Auguste Comte, "for when you're home on leave"; to me these plans seemed somewhat unreal.

On Monday I left for Cherbourg with Eveline, who had decided to accompany me. All she had with her was what she had needed for one night in Rouen, and I had scarcely more than she. Never had we traveled so light! At a station along the way, there was a canteen serving soldiers for free. Equipped with my induction orders, I was now a soldier, and my first act as such was to order coffee for my wife and myself.

In Cherbourg, I was disguised as a soldier. Except for the color, I saw virtually no difference between this blue uniform and the prison garb that I had just doffed in Rouen. The lieutenant under whom I was to serve was a Protestant pastor in civilian life. Having already familiarized himself with my case, he gave me a moralistic sermon, exhorted me to expiate my past errors and do my duty bravely, and informed me that the Germans had just made a breakthrough in the Ardennes.

From my whole stint in Normandy, there are only a few clear memories still afloat in my mind. In Cherbourg, my company was detailed exclusively to the task of going to the train station every day to load shells on trains heading north. I told my comrades: "Don't wear yourselves out, these shells are going to end up in your buddies' faces anyway." My reasoning was off by only one detail: the Germans were too well equipped, and in too much of a hurry besides, to be tempted to use cannons and ammunition seized from the enemy.

Eveline's hotel was along the route my company took every day. Once, after arranging it with my comrades (who were only too delighted with the escapade), I slipped away to spend the day in Eveline's room, resuming my place among the troops on the return march. Another day (a Sunday, I suppose), the two of us enjoyed a picnic of cold lobster and a bottle of Mercurey, on a hill overlooking the naval base. Never has a Mercurey tasted so good to me.

Soon it was time for Eveline to leave again. With the Germans approaching, Cherbourg was becoming a restricted army zone; no one could take the train without special authorization. The trains and roads were streaming with refugees from all directions. Eveline had to return to her mother and Alain in Parcé, where she arrived only after a long and uncomfortable trip, including a whole night spent in the train station in Le Mans.

At the beginning of June I was transferred to Saint-Vaast-la-Hougue on the coast of the Cotentin peninsula. I was assigned to a machine-gun company. Some of the weapons were positioned for anti-aircraft use. From time to time, when one saw, or thought one saw, a German plane, one would fire a few rounds. Once the machine-gunner claimed to have hit his target. If so, this was the only time I took part in actual bloodshed: I had helped carry the ammunition chest.

These inoffensive tasks left me with ample free time. I took sun-baths by a creek, where I would read Lefschetz's book on analysis situs and algebraic geometry – the book about which Hodge used to say that all the important statements were true and all the others false.

At that time wild rumors about German paratroops were frequently heard. The story about parachutists disguised as nuns, hiding a motorbike underneath their skirts, is still remembered. It was also said that the parachutists wore spring-soled boots, supposedly to absorb the shock of landing. One day a villager came to report to the captain of my company that he had seen a paratrooper come down in the parish priest's garden. When he landed on his spring-soled boots he bounced back up again and rose into the sky.

For several days, we saw English reinforcement troops disembarking and passing through the village on their way to the combat zone. The weather was by then quite hot, and village women were offering water to the English soldier. I saw one sergeant recoil and dash the water on the ground. He had probably served in the colonies and had got the idea that all "native" water was dangerous to drink.

We were given little information on the military situation, save for sketchy reports on French radio, but it was all too obvious that it was deteriorating rapidly. One day a villager, ready to leave for the Bordeaux region with his family, asked me if I thought they would be safe there. Without hesitation, I replied in the affirmative. It was difficult to let go of the idea – a holdover from 1914 – that the front line would have to stabilize somewhere, probably at the Loire river.

One day, we saw thick black clouds in the sky. Perhaps there was nothing extraordinary about them, but in no time there was a rumor that Rouen was burning, and that these black clouds were from flaming oil tankers. I never found out if there was any truth to this rumor, nor how it had made its way to us.

On June 17, the regiment moved to a military installation at the water's edge, inside an old enclosure with seventeenth-century fortifications *à la* Vauban. It was said that the Germans were nearing. Loudspeakers broadcast Pétain's speech announcing that he had called for an armistice. I was not at all sorry; from the little I had been able to observe of the general incompetence and demoralization, and from what I had seen and heard of the masses of refugees, it seemed obvious to me that this was the only possible way out. Moreover, I still had an unshakeable confidence in the English, though I would have been hard put to explain it. The regiment took up a defense position; for a moment I had the impression that, out of some misplaced sense of professional duty, our officers were leading us into a "glorious," but utterly pointless, death. I was probably wrong, but I told a few fellow soldiers my opinion. Our pastor-lieutenant, who was always poking his nose into everything, heard me say it, and immediately had me locked up in a sort of shed.

I was in the process of studying possible means of escape when the door opened. One of my comrades explained that the order to evacuate had just been given; the lieutenant was ready to abandon me, but the soldiers had protested. And so I rejoined my company.

Thereupon the order was given to deploy a section of machine guns for action under the command of a sergeant, to cover the retreat with heavy fire. The German tanks were not yet in sight, but we knew they were not far away; German planes were flying all over the sky, completely unopposed. I asked the sergeant – a reservist, not a professional – , "So we're to be the casualties?" He assured me that he had no intention of getting us killed. Just then came the command to abandon our machine-guns and join our regiment on the beach. I was later told that, strictly from the perspective of our safety, this command was an egregious error on the part of our officers: machine-gun fire can discourage enemy planes from getting too close. Whatever the case may be, we found ourselves on the beach. A German plane, flying quite low, dropped a bomb some 300 yards away from us, clearly not intending to harm us. We were boarded on a small steamship which appeared to be waiting for us, and which put out to sea without being attacked. My comrades and I were in the dark as to our destination; our officers were perhaps no better informed. There was talk of Brest, but when we disembarked the next morning, we were in Plymouth.

Since leaving Cherbourg, I had stopped shaving. After landing in England my first act, as soon as it was possible to wash up, was to shave my nascent beard – but I spared my budding mustache. Asked by an English civilian whether the war was lost, I assured him of my utter confidence in England. We were boarded on a train; how long the trip lasted I do not recall. When we arrived our regiment was added to a camp already full of French troops. This camp had been set up in central England, not far from Stoke-on-Trent, primarily for the purpose of housing troops back from Norway. I thought I understood it to be under the command of a General Pétoire, a name which somehow seemed symbolic to me;[1] in fact, it was a certain General Béthouart.

It was just after the 18th of June, which was when de Gaulle launched his appeal to the French. The officers and even the soldiers in the camp had a decision to make: whether or not to join forces with the Gaullists. It was rumored that an officer who joined de Gaulle's cause stood a good chance of rapid promotion, especially if he brought his troops along with him. The commander of a Moroccan company summoned his men to the tent that served as his "office" and had them march one by one to where a sergeant had them sign a paper. Afterwards, they were told they had just enlisted in the Gaullist army. I felt no urge to follow suit. Nevertheless, the situation had changed greatly since the preceding year. I wrote to colleagues in Cambridge that, if I was to be unable to devote myself exclusively to mathematics, I would not be averse to putting my scientific knowledge to use in the service of the English, if there was any way in which I could be useful.

We were at loose ends in this camp. Within certain hours, we were allowed to go to town, where we drank beer, chatted with the English (both civilians and military), and – in my case – bought the *Times*. Some comrades and I extended these hours by jumping over the walls. I was caught. In punishment for this misconduct my pastor-lieutenant sentenced me to a few days in "prison": prison in the military sense, of course, which in a garrison means a stay in the disciplinary facilities. In the camp, one was confined to some tents surrounded by barbed wire and guarded by a company of Gardes Mobiles.[2]

Then came orders to leave the camp. The entire camp, that is everyone who was not enlisting with de Gaulle, was to board ship at Bristol, heading for Morocco.

I was still in the custody of the Mobiles, who for some reason unknown to me were delayed along the way. When they made it to the quay, the ships they were supposed to board had just pulled away from shore. I stayed in Bristol with the Mobiles. My first reaction was, foolishly, to be annoyed. In fact, it turned out to be a stroke of luck: I was thenceforth definitively parted from my conduct sheet. Who knows what fate would have befallen me if I had embarked with my company, especially since the incident of the day before leaving for England – which my pastor-

---

1   *Translator's note:* The word *pétoire* is French for "peashooter."
2   *Translator's note:* The Gardes Mobiles were a certain section of the gendarmerie, known in brief as simply the "Mobiles."

lieutenant had pretended to forget – could well have reared its head in the future? The following winter, when I was back in France, I learned that there had been rumors to the effect that I was breaking stones in a penal quarry in Tataouine. That scenario was by no means impossible.

The English put up the Mobiles in wooden barracks by a pretty village near Bristol. No longer a prisoner under their guard, I had been promoted to company interpreter, a role that I found congenial. Besides the Mobiles, there was a company of former Spanish Republican soldiers whom the French had drafted into the Foreign Legion. The English hadn't the foggiest idea what to do with them. An English commander was in charge of the whole lot of us. I took great pleasure in offering my interpreting services to the Spaniards, who were living rather cheerlessly between their barracks and a wasteland which they used for exercise. They could see the village soccer field not far away. At their suggestion, I asked the commander to let them use it, and he consented. After a few days, the local team saw fit to propose a match, which the Spanish won easily. Their pride wounded, the villagers rallied their strongest team – only to be defeated once again by the Spanish. From that time on the Spanish, who had until then been the object of distrust, became quite popular. They now had free run of the village and vicinity. In gratitude, they invited me to a party where the star attraction was a talented flamenco singer, an eighteen-year-old who had already served four years as a soldier. Shortly thereafter, the Spanish received a visit from Michael Foot, the Labor M.P. I do not know what became of them afterwards.

As for my Gardes Mobiles, the presence of the Spaniards gave them an opportunity to recall, with loud guffaws, the time when they had been detailed to guard a refugee camp for Spanish Republicans in the Pyrenees. Their talk left no doubt as to the rough, even brutal, treatment they had inflicted on these unfortunates. When the Gardes Mobiles called on my services, it was sometimes for rather strange purposes. One day they wanted to make it known in no uncertain terms that, as they were non-commissioned officers, they were exempt from the obligation to make their beds and sweep their quarters. They insisted that I transmit this message to the English commander. His reply was that if they didn't sweep their rooms, their rooms would not be swept.

Fortunately, my duties left me free a great deal of the time. I went regularly to the public library and took many walks in the area; I was even free to go to Bristol, where I made contact with colleagues at the university. A physicist, about to be mobilized in the air force, told me despairingly: "I am certain I will be ordered to bomb the cathedral at Chartres, and I am certain I will miss it."

One morning, the English radio announced the bloody Mers-el-Kebir affair, where the English fleet, thinking it necessary to incapacitate the French fleet seeking refuge near Oran, suffered severe losses. First I thought that it would be wise not to show my French soldier's uniform in the streets; then curiosity got the better of me. I have never had so much admiration for the English as on that day. I heard only expressions of understanding, even of sympathy. Only once did a man, who was clearly under the influence of alcohol, lash out at me verbally; his companions shut him up, telling him, as I myself could have done, that when they fired on the English the French navy thought they were doing their duty, and that in any case I was in no way responsible.

One morning toward the middle of July, I was told to pack up my belongings – a task that was finished in short order, as all I had was my knapsack. At the end of the trip I found myself surrounded by barbed wire, in the company of French soldiers coming from various directions. Some of them were legionnaires who had fought at Narvik in Norway. We were housed in buildings that had previously sheltered captured German officers. There were even showers for us to use.

At the evening roll-call we fell into line and were reviewed by a young lieutenant escorted by a non-com who looked like a prison guard (and had probably been one in civilian life). When it came my turn, I did not stand to attention. I believe this was sheer negligence on my part; but when the sergeant barked at me to salute, I didn't budge, and he pounced on me. I think he and the officer thought this was the germ of a revolt. I was taken off to a small shed and locked up. A little later, when it was clear that I would be left there for the night, some comrades came with blankets for me – the nights were cold there.

No doubt I would not have behaved this way if I had been dealing with anybody but the English, but I thought I knew these people, and as it turned out I was not mistaken. In the morning, I was taken to the camp commander, an English colonel, and invited to speak in my own defense. I said that to my knowledge, we as French soldiers were still entitled to be treated as allied soldiers by the English; a salute is due to officers of an allied army only as a courtesy, but cannot be enforced as a matter of military discipline. I was ready to salute the English officers out of courtesy, but not on command. If, on the contrary, the English considered us to be prisoners of war, then we had no choice but to submit to their commands: this, I said, was precisely the point that my comrades and I wished to have clarified. Moreover I assured him, quite truthfully, of my pro-English sentiments. My speech could not have been more successful. The colonel admitted that he himself did not know what our status was, and he wanted nothing so much as to see this point cleared up as soon as possible. He sent me back to my comrades, exhorting me to encourage their cooperation.

My stint in this camp lasted approximately two weeks. We organized a modest program of mutual instruction, including English courses provided by me. My method consisted in imagining the conversation that one of us might have with a young Englishwoman met outside the camp. I analyzed the conversation sentence by sentence, gradually introducing the most common verbs with their conjugations. It seemed to me that my students made real progress within two weeks. Some of them, knowing that I had not a penny to my name, expressed their gratitude by buying me chocolate bars at the camp canteen.

One of my comrades, a small electrical engineer, was a Jew of Polish origin who had lived in Germany for some time. Fleeing Hitler, he had emigrated to Paris, where he became a simple laborer. He told me how, one or two years before the war, an officer from the French intelligence service had tried to recruit him to go "work" in Germany. Everything had been arranged for him: a position as an engineer, housing, strictly "Aryan" identity papers; all he would have to do would be to tell a visitor from time to time – a different person each time – what he had seen in his factory. According to the officer, there was absolutely no risk involved. My friend objected that his looks and accent would instantly give him away as a Polish Jew. The officer insisted that he would be provided with faultlessly counterfeited papers, and all would be perfectly in order. After this fruitless exchange, which lasted for some time, the officer still did not appear to understand. When war broke out, my friend was drafted into the Foreign Legion and served as an interpreter during the Narvik expedition. It must be from him (unless it was from someone at the camp near Stoke-on-Trent) that I received one of the few war trophies taken from the enemy by the French army in 1940: a German military-issue blanket brought back from Narvik. He asserted that the success (local though it was) of the French army on the Narvik expedition was entirely due to the Spanish legionnaires.

It was not without regrets that, toward the end of July, I left my friends and pupils, both French and other nationalities, and moved to the White City camp in London. This was a greyhound racetrack where the English had housed some 2000 French soldiers, almost all of them recently discharged from English hospitals. For the most part they were survivors either of the Norwegian campaign or of Dunkirk. Having been treated in England for wounds or illness, they were now either cured or in the process of recovery. Here I was given a mattress and a blanket. Most of my new comrades were scattered here and there underneath the tiered bleachers in the stadium. Some, myself included, had chosen a place on top of the bleachers: here we were shielded from the elements and from exploding anti-aircraft shells, but still in the open air. The only problem with this arrangement was that every Saturday there was a greyhound race, and we had to clear out for the benefit of the spectators. It is true that a section was reserved for us so that we too could watch the races. These were discontinued when bombs dropped on another stadium, in a London suburb, claimed victims – not only among the spectators (something the English would have easily accepted) but among the greyhounds: this was going too far.

We were entirely at liberty to roam about London all afternoon. In fact, we even had our evenings free, as jumping over the wall couldn't have been easier, and the English winked at those who did it. The frequency of anti-aircraft fire, especially in the evening, made it advisable to wear our helmets. At one point, the English deemed it prudent to equip themselves with gas-masks. Some of my companions, worried because they were not issued any, asked me to petition the authorities on their behalf. The camp commander told me, "They can have mine if they wish."

In their own way, the English did show concern for our fate. One day, we fell into rank for review by the Queen and Mrs. Churchill, who shook a great many hands. A rather elderly Francophile invited us out in small groups, to boost our morale. One day he took me to a tearoom with several other soldiers, good French country boys who no doubt would have preferred an aperitif to tea. They were already intimidated enough by the setting; on top of this, my Englishman asked them: "China or Ceylon?," and endeavored to convince them that the two were as different as red and white wine. As far as I know, the invitation was not repeated.

There were all sorts of rumors as to what was to become of us. In July a French hospital ship, which had come to pick up wounded and ill soldiers, had been sunk in the English Channel as a result of confusion between the English and German commands. Despite the advice of some Frenchmen that the affair was best forgotten, English radio broadcasts in French were constantly referring to this lamentable mishap in an effort to shift the blame away from the English. As for us, it was rumored from time to time that we were soon to be repatriated, but nothing seemed to come of these rumors. One day it occurred to me to measure the speed with which such rumors were propagated. All I had to do was to start one – the more preposterous the better – at one end of the stadium, and then hasten to the other end to await the results. Unfortunately, I left London before I had a chance to carry out this sophisticated sociological experiment.

The Gaullist propaganda services sought to take advantage of our ambiguous situation by attempting to recruit from our ranks. They met with little success; my companions, primarily country boys, wanted only one thing: to return home as soon as possible and be reunited with their families, with whom they were completely out of touch. The masterpiece of Gaullist propaganda was to have us assemble one day inside the stadium and to announce through the loudspeakers, in a tone of forced sadness, that we were all to be shot upon returning to France. The grain of truth behind this assertion was Pétain's decree that made all those who had voluntarily emigrated to England to join the Gaullists liable to the death penalty. But even the most benighted of our country-boys could not fail to understand that if Pétain was sending a boat to take them home, it was not to shoot them on arrival. One might object that later on Stalin did just that; but Pétain was no Stalin. I heard other echos with respect to the Gaullists' methods of recruitment, and all of them left me with a very dim view of this movement, and even of the man who encouraged or condoned these tactics. I was perhaps the only Frenchman in England at that time who was both pro-English and anti-Gaullist.

As soon as I had arrived in London, I had tried to contact friends. A German family, longtime friends of my parents, lent me some money. While all mail service to occupied France was suspended, it was possible to communicate, preferably by telegram, with the so-called "free" zone. Through Henri Cartan, who had remained in Clermont-Ferrand with the University of Strasbourg, I was able to communicate with my parents and my sister, whose address in Vichy I obtained. In London itself I located my friend Frank Smithies, from Cambridge, who worked in a technical ministry, as well as my friend and colleague from Strasbourg, the chemist Guéron, who had enlisted with de Gaulle's forces and who was later to work on developing the atomic bomb, in Canada. His family had stayed behind in France, but was able to join him several years later. I also went to see my friends the Brahams, whose only son I had gotten to know in Cambridge. This brilliant student, mobilized as a pilot, had just been reported missing in action. His parents, who held out no hope of seeing their son again, bore their grief valiantly; I cannot find words for the warmth with which they welcomed me. On my first visit, they asked what they could do for me. My answer was: "Allow me to take a bath." Then they fed me dinner. They lived in an apartment in a large complex of terrace-roofed new buildings in Chelsea. On one of the evenings I spent with them, the air-raid alarm sounded toward the end of the meal. They went down to the basement, and I asked to go up to the rooftop. They voiced no objections; only Mr. Braham said to me, as I left them, "If need be, would you prefer to be buried or cremated?" I did not regret my curiosity on that occasion. All of London and its suburbs were the scene of a stunningly beautiful barrage of anti-aircraft fire. Tracer bullets were streaming in every direction; on all sides, searchlights swept across the sky, trying to catch enemy planes in the net of their beams. By their light the

London (August, 1940): A.W., seventh from left (with mustache)

tethered balloons shone with a sharp metallic glare, forming a bright ring around the city. No fireworks display could compare to such a spectacle, which I later found described in Saint-Exupéry's *Flight to Arras*. As his navigator remarked, "You wouldn't see that in civilian life."

In the month of August, German airplanes made almost nightly visits. These planes had two motors which, for reasons unknown to me, were out of phase, and the pulsing produced a characteristic whirring effect. I am sure no one who heard it then has been able to forget it. As loud as the anti-aircraft fire was, I had grown accustomed to it, and it did not keep me from sleeping. One evening in particular, when I was already in bed, I heard the air-raid warning, then the whirring of a German plane, which was quickly drowned out by the thunder of anti-aircraft fire. The latter would stop momentarily, making the whirring intermittently audible. I thought of Richard Strauss's *Till Eulenspiegel,* where the light-hearted, mocking Till motif recurs intermittently amid the ruckus of forces opposing him. I took an interest in the fate of the German bomber. Each time his presence was audible again, I was thrilled: it seemed to me a victory of mind over matter.

I walked far and wide throughout London. Never have I had another chance to spend such a long time there. I visited its many lovely parks, both large and small. I also went to the National Gallery, and sometimes to concerts which were held from noon to one o'clock to compensate for the lack of evening performances. From time to time I went to the French consulate to glean some clue as to the fate of the inhabitants of White City. I was courteously received, but came away with nothing; perhaps they knew no more than we did.

Our Frenchmen would hardly have been worthy of the name if they had not had their escapades with women. To the exalted reputation of the French in amorous affairs and the relative shortage of males was added the allure of men in uniform, with the result that conquests were not difficult. Moreover, they were free, for it was no secret that we had little or no money. The farm boys, who made up the great majority at White City, were generally peaceful, hardly leaving the camp as a rule, but there was no shortage of city boys looking for any opportunities that might present themselves. The minute they left the camp, there were women to meet them. Some of them put on a show of battlewise braggadocio: when the alarm sounded and the women wanted to head for the shelter, these blusterers would refuse, saying that they had been through it a hundred times already. Later, when the bombing grew heavier, these brave boys (at least, according to confidences I heard from them) changed their tune: this time, they were the ones who wanted to take shelter, and the women would be saying, "But darling, you've taught me not to be scared!"

Of course these contacts were not always risk-free. One of us, appearing before the doctor one day, was told: "Well, old chap, you've got the clap." "Jeez," he replied, "I never would have expected it of an English girl."

In September the bombing got heavier, first in the poor East End neighborhoods. With a friend I went to survey the damage. Though infinitely worse was still to come, this was already a wretched sight to see. When the so-called London Blitz took place, I happened to be taking a walk in Richmond Park. That afternoon Göring had unleashed his huge armada over London. Thanks to the English radar, it suffered such damage that he didn't try it again. It was found out later that a large number of his planes had been stopped before getting to London, but enough were left to create quite a spectacle. Lazily sprawled out on the grass in the park, I saw the whole sky filled with planes chasing each other every which way. From time to time, one of them would spiral down out of the sky; sometimes I could see a parachutist bailing out. That evening, there was not a bus or a boat to be found from Richmond back to the city. I walked all the way back, following the banks of the Thames. It was a marvelously calm and peaceful night.

For those of us at White City, the rumors of departure seemed to grow more precise. Churchill gave his famous speech, saying "We shall fight on the seas and oceans... We shall fight on the beaches..." Hearing him on the radio I became convinced that the invasion had already started. We were informed that we would be leaving London. Did this mean we would embark for France? It seemed equally plausible that we would be sent to a concentration camp. Despite the insistent advice of French and English friends, I had resolved that if my companions were repatriated I would go with them. I planned to send for Eveline and Alain on reaching France, and then to leave with them for the United States. I saw no way of staying in England unless I enlisted with de Gaulle. Even if I had been able to accept this idea, I could hardly have hoped in that case to have Eveline join me, or even to communicate with her effectively: as an occupied zone, Parcé was cut off from all correspondence with England. Besides, my curiosity had been whetted. In England we had no more of an idea what was happening in France than if we had been on another planet. Neither reading the newspapers nor conversing with friends shed any light on the subject. In short, I was dying to see what was going on in France.

I had absolutely no identity papers. To be ready for a possible departure, I had a comrade in the camp office prepare a card using my real name, certifying that I was just recovering from pneumonia, and naming the hospital where I had been treated. With this card, I had no difficulty boarding the train for Liverpool when the time came to leave. At Liverpool there were two fine hospital ships, the *Sphinx* and the *Canada,* which Pétain had sent over for us. I was truly sad to find myself separated from the best friend I had made at White City.

This is the closest thing to a cruise I have taken in my life. The cabins were reserved for officers and for the seriously wounded or gravely ill; but I had no trouble finding myself an airy spot. There was plenty of food. Because of my so-called pneumonia, I was X-rayed; no lesions were found.

The *Canada,* with all her lights showing and a large white cross painted on her top deck, followed a carefully plotted itinerary that had been agreed upon by the English, the French, and the Germans. First we headed north. Off the coast of Ireland we witnessed the rescue of two English aviators whose plane had been shot down, and who were floating on their life raft. They were not badly hurt, I believe, but were nonetheless gladdened by the prospect of being attended by real physicians. After completely rounding Ireland, the ship steered a course south through Portuguese and then Spanish waters, all the way to Gibraltar where the two Englishmen were dropped off. After putting in briefly at Oran, the ship headed once again toward Spanish waters, and finally made for Marseilles

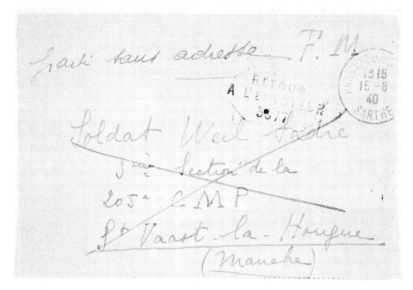

Eveline Weil's last wartime letter to A.W. (June, 1940)

where we docked after two weeks on board. The bulk of my time had been spent sunbathing.

I believed my parents and sister were in Toulouse, where they had indeed gone after a time in Vichy. In London, and again in Oran, I had tried to contact them to warn them of my arrival on the *Canada*. When the boat had docked, I was surprised to see them waiting for me. They shouted to me, "Oscar is here waiting for you." "Oscar" was our code word for the police.

I was later able to reconstruct what had happened. My mother, worried as always, had been badgering any number of people to find out if I was really on board. Of course she had not been able to learn anything, but while waiting on the quay she had heard some officer, police agent, or bureaucrat say, "We are also looking for a man named André Weil." Not realizing he was inquiring about me for her sake, she mistook him for "Oscar." At her warning, I went back to my modest baggage and made haste to destroy a letter entrusted to me by a friend for his family. I do not know how important this message was to him; he never did find out why his family did not receive it.

Nothing happened. We were taken to barracks and given a few hours of freedom, which I spent with my family. The next day, upon showing the card fabricated in London, I was duly discharged. Like all my

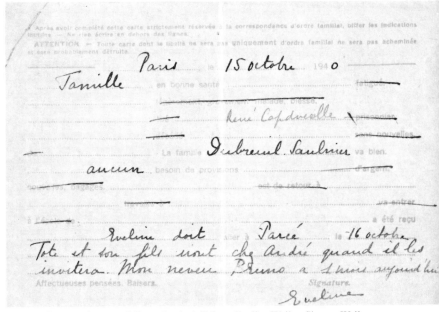

Interzonal postcard ("carte bochtale") from Eveline Weil to Simone Weil
(October, 1940)

companions, I was issued the suit made of coarse blue military fabric, cut
in a civilian style, to which every discharged soldier was entitled. It was
known as the "Pétain." I had to state my profession. Already the previous
winter, even before I had returned to France, and even before any precise
information on me could have found its way to France, I had been
dismissed from my Strasbourg position – on Daladier's personal order,
according to what I was later told. Nevertheless, I said "professor," which
seemed to imply that I was an employee of the state. The anti-Jewish laws,
by virtue of which I would have been dismissed anyway soon afterwards,
had not yet been announced. As a consequence, I was not entitled to the
small sum of money granted each discharged soldier to aid in the return to
civilian life. I did not raise any objections. As he held out the "Pétain" suit
to me, the elderly officer presiding over the ceremony told me, "You can
use it to go fishing."

A Farewell to Arms

I was a civilian again, and an unemployed one. My parents assumed the
burden of my financial problems. They were not in the habit of discussing

their financial affairs with me; I accepted their help without asking questions, hoping of course to be some day in a position to repay them.

I had had no news of Eveline since June. The last letter she had written that month, using the correct military address for me, had come back to her in July stamped "Relocated, no forwarding address." In truth, the entire French army had relocated, leaving no address. I still have this symbolic document, which I save for future historians. Eveline had somehow found out that I was in England. In September, a local water diviner, who was also the postman in Brûlon, confirmed this fact and predicted my return – even, apparently, describing approximately the route I would take. Equipped with a map of Europe and his divining-rod, he would locate missing persons using a photograph and the individual's last known location. Eveline had mentioned only Cherbourg, but apparently the divining-rod traced my path through England and then to the Mediterranean. How much of this is to be believed? Unfortunately, no transcript of this consultation exists.

At the beginning of October, correspondence between "free France" and "occupied France" had just been authorized, but was limited to "interzonal family cards," which Henri Cartan dubbed *"cartes bochtales"*.[1] These were pre-printed cards, offering a choice of formulaic messages pertinent to French life: everything from "-- is in good health/has been killed/wounded/taken prisoner" to "-- has passed/failed the -- exam," followed by two lines for "personal correspondence" and then a choice of two closing formulae, "Love and kisses/Affectionately." These cards had only recently been distributed, and so far none of them had reached its destination.

In addition to the problem of communicating with Eveline, I was preoccupied with that of finding a way to leave for America – legally, if at all possible. Nothing appeared easy: passports, visas, transportation. From the start of the war, the University of Strasbourg had been relocated in Clermont. Though I no longer had any official tie to it, that is where my colleagues were; moreover Clermont was close to Vichy and thus not far from the demarcation line between the free and occupied zones of France. This was the line that Eveline would in any event have to cross in order to join me, and which I was ready to cross myself if need be, despite the obvious risk that would entail. I sent a telegram to Henri Cartan and took the train to Clermont.

1 *Translator's note: "Boche"* is a derogatory term for Germans, the French equivalent of "Kraut"; the expression *"cartes bochtales"* echoes *"cartes postales,"* the French for "postcards."

I did not think that "Oscar" posed a serious threat; nevertheless, my sister had not been able to determine precisely the import of the suspended sentence I had been granted. It was not wholly out of the question that I still had to "serve time", and in the political climate of the day a denunciation could have had unpleasant consequences. As soon as I got off the train in Clermont, Cartan said, "Watch out, you might be spotted." I explained my tactics to him: I had decided to let myself be seen everywhere exuding a sense of utmost security. If I appeared to be hiding, that could suggest to some charitable would-be informer the idea of a discreet denunciation. This was a risk I didn't wish to take. I went to visit all my friends. Although Dean Danjon was no longer "my" dean, he assured me of his support, and kept his word: he personally escorted me to the Ministry of Education in Vichy.

At this time, and even more so once winter arrived, Vichy was a curious sight to see. The major hotels had been converted into government offices, with the beds removed but the wash-basins and bidets intact. As the central heating was not functioning, stoves had been installed, with their exhaust pipes fitted through windowpanes specially cut out so as to accomodate them. All these pipes poking out gave the façades a bizarre, bristly look from the outside.

Thanks to Danjon, no doubt, I was well received at the Ministry. The Vichy men were for the most part people of good will, sometimes even Gaullists. In any case, they were not displeased to dispatch cases like mine by exporting them to America. Terracher, Rector of the University of Strasbourg and also Secretary General of the Ministry of Education, was helpful and encouraging. I was also received by Belin, no doubt upon the recommendation of my sister, who had corresponded with him for some time before the war. She thought highly of this trade-unionist, an honest and very worthy man whose only mistake was in thinking that he could accomplish some good as part of Pétain's cabinet. Belin introduced me to one of his co-workers thus: "This is so-and-so, a pacifist; this is André Weil, conscientious objector." He promised to back my request for a passport. I do not know whether he ever did have occasion to intervene on my behalf.

Eveline's brother, René Gillet, was then in Vichy with his wife and his son, a newborn named Bruno. Vichy is not very large; I ran into René by chance at the park. A lieutenant in active service, he was due to leave for Syria before long. Somewhat later I received a telegram from my parents: "Nephew Bruno is one month old." They had just received Eveline's first interzonal postcard. Not knowing what to put on the "personal message" lines, she had written, "My nephew Bruno is one month old"; my parents, convinced this was an urgent message in code, wasted no time in passing it on to me.

Apart from my personal problems, there was Bourbaki's survival to think about. There had been no meetings since the congress in Dieulefit in September of 1938. The first issue of our internal bulletin, "The Tribe," had been put together and distributed by Dieudonné in March of 1940, and so far no subsequent issues had appeared. A first installment of our great work had already been published by Freymann. Fully aware of the role we expected it to play, and in memory of Euclid, we had called it *Elements of Mathematics*. A second part had just been, or was about to be, published. In autumn of 1940, we needed to take stock and rally our group, insofar as circumstances would allow, with a view to continuing our work. Fortunately, several of us were now in Clermont. Delsarte, not allowed to return to Nancy (which was in the "forbidden zone"), was in Grenoble filling in for Favard, a prisoner of war. We decided to hold a congress, adding to our ranks Laurent Schwartz, a brilliant young mathematician living in the area. This plan almost fell through when Delsarte sent me a telegram saying "OK for congress in Clermont." The word "congress," with its political connotations, was taboo in Vichy France. I was summoned to the regional military command, where I went thinking it was something having to do with my passport requests, which at that time had to be approved by military authorities. An officer asked me, "Do you know someone by the name of Delsarte?" When I replied in the affirmative, he asked me to explain the telegram. Memories of Finland came back to me. Nevertheless, after I had answered and called upon Danjon to vouch for me, I was courteously dismissed, and our "congress" took place as planned.

The most pressing of my concerns, though, was to communicate with Eveline. The interzonal postcards were woefully inadequate, and other means of communication were slow and unreliable. I went to Vierzon, which straddled the demarcation line. The checkpoint was a bridge guarded by German soldiers. The bridge saw a brisk traffic, both legal and illegal. On the free side, I met the mayor, who assured me that my problem was quite simple. He offered to take a letter to Eveline, promising to mail it on the other side. Following his instructions, I asked Eveline to report to a certain address in Vierzon, whence she would be taken to the free zone. On receiving my letter, Eveline went to Paris, taking Alain along, of course, and also her mother, who wished to see her son René and his wife again, and the famous Bruno. In Paris, they were given contradictory advice, and even told that they should contact the *Kommandantur*. From there they went to Vierzon. But by then the address I had given her was no longer any good; the border escort must have been caught. Fortunately, at Vierzon the line was guarded only by the regular German army: a little later, the SS took over, with much harsher control. Anyone crossing the line illegally might be shot, but was more likely to be sent to a concentration camp for an indefinite period of time. This was enough to make one hesitate. However, there was no lack of smugglers, as long as one was willing to pay the going rates. At the critical moment, a peasant girl commanded the attention of the non-com on duty, and Eveline, her mother, and Alain were able to cross – if not without qualms, at least without mishap.

I received a telegram, sent from Issoudun, announcing Eveline's arrival in the free zone. I went to meet her and found her in some railway station where she had to change trains. We had not seen each other since Cherbourg. We stayed in Ceyrat, not far from Clermont, in a hotel where provisions were still plentiful – unlike the city, where pickings were already slim. On the other hand, this was a severe winter, and the only heat was from stoves that had to be put out at night. Early in the morning I would force myself to get up and light ours. When I was unsuccessful (which was fairly often) the whole room would fill up with smoke and we would have to wait for a more expert maid to rescue us. Of course we had our breakfast in bed, snuggling under all the woollens we could get our hands on. Dieudonné, staying in the next room, made an outlandish sight in the early morning darkness: swaddled in a staggering number of sweaters, he also wore a knitted night-cap of thick wool which covered all but his eyes, mouth, and nose. But the walks we took in the snowy woods quite near the hotel more than compensated for these minor discomforts.

Meanwhile, I had contacted various American colleagues as soon as I returned to France. I received a cable offering me a teaching position at the New School for Social Research, and recommending that I apply for a "non-quota" visa at the nearest American Consulate; this was in Lyon. I had heard of the New School and knew it to be a liberal institution, founded fairly recently and devoted principally to the social sciences. I did not think mathematics was taught there, but this was hardly a time to quibble. At the consulate, however, I was unpleasantly surprised to learn that, according to American law, a "non-quota" visa could not be granted unless the applicant could show proof of a currently held teaching position. In January I had been dismissed from my position at the University of Strasbourg. The consul, whose sympathy for all the masses of Jews desperately seeking visas was clearly in short supply, rejected my application on these grounds – or this pretext. In the meantime, Eveline had joined me, and we had received our passports, duly approved by the Vichy authorities. In Marseilles, my parents had succeeded in booking passage for Eveline, Alain and me on a ship sailing for Martinique. I cabled the New School, and was instructed to leave for Martinique; in the meantime the school would see what could be done about obtaining an American visa, for which they would no doubt lobby in Washington. And so we left for Marseilles.

Vichy had just passed the first anti-Semitic laws, and I had already seen them enforced in Clermont, at the University. The deans of all French universities were required to list all Jewish faculty members teaching in their institutions, as the first step in forcing them to resign. Many deans found a way out by asking all their colleagues to designate themselves as Jewish or non-Jewish. This was a decent approach; still, I have always considered it a sad reflection on the honor of the French universities that not one dean took the stand of saying, "I have never known which of you are Jewish and which are not, and I do not wish to know. I hereby submit my resignation as dean; let someone else perform this task, if he will." I do not believe that a dean who chose this course of action would have incurred any real risk. As Shafarevitch once said, under such a regime there are many circumstances when one can, without any real risk, either straighten up somewhat or bow down a little more. In any case, under the Vichy regime, even those who lost their teaching position for racial reasons remained personally safe; this was to change in 1942, with the advent of total occupation. In the meantime, Marseilles was already full of would-be emigrants, not all of whom were Jews or foreigners. They crowded the waiting rooms not only of the United States consulate but of many others besides. Somewhat later, a reputable Marseilles businessman, who happened to be the honorary consul for Siam, hatched the ingenious plan of selling (at a rather steep price) perfectly authentic visas for Siam. I still have my parents' passport stamped with one of these visas, which might as well have been a visa for the moon. A number of refugees had pooled their resources to rent a villa, which they baptized *Château Espère-Visa.*

I too – though not very hopefully – went to the American Consulate in Marseilles. This time I had better luck. The consul was a man whom Roosevelt had sent over with the express mission of saving European intellectuals endangered by the war. At my first request, and no doubt in defiance of all the regulations, he granted me "quota" visas.

Despite the pessimistic predictions of several of our friends, our preparations were thus complete. Our passage was booked on the *Winnipeg,* which along with her sister ship the *Wisconsin* constituted the entire fleet of the proud Compagnie Générale Translatlantique (popularly known as the "Transat"). The *Winnipeg* and the *Wisconsin* were two mixed-cargo (passenger and freight) "banana boats" plying trade with Martinique. Eveline announced her imminent departure to her mother (who had already returned to Parcé) using a *carte bochtale:* "Alain is starting nautical school." In return, she received a card saying, "André's children are at Uncle Henry's." These "children" were my books. In Paris on her way to join me, Eveline had arranged for my library to be transported to Enrique Freymann's, with the help of my parents' faithful cleaning woman. At about the same time, Freymann sent me the first two copies of my book on integration in topological groups, which I was thus able to take with me to the United States.

And so, at the end of January, we boarded the *Winnipeg,* equipped with all the sacraments and a large bucket of jam – which was not actually jam. During our stay in Clermont, the food shortages, although not dire, were beginning to be felt. Everyone was always on the lookout for a good deal. At this time my mother-in-law was taking it upon herself to send her Parisian friends packages of foodstuffs, as she would tell us by *carte bochtale:* "The Colidan-Rée[1] family is doing well." In Clermont one day, word got out that the famous preserves factory "Marquise de Sévigné" was selling second-quality candied fruit. Our wait in line (not overly long) was rewarded with a bucket of delicious fruit preserved in syrup, a significant addition to our diet during the voyage of several weeks, for the ship observed the rations decreed by Vichy law – meaning that the less fore-sighted or less fortunate passengers were reduced to dry bread for break-fast. A facetious passenger remarked that the Transat's new motto was, "Bring your own food." On the other hand, this ship which had by force of circumstance become the flower of the Transat fleet was staffed with the line's top personnel. In the dining room, we were greeted by the principal *maître d'hôtel* of the *Normandie.* Every evening, no matter how unbear-ably hot, he insisted on jacket-and-tie dress for dinner. No doubt he suffered because he could not insist on tuxedos.

This was not the only exceptional aspect of our crossing. We were sailing under the yellow flag of the Armistice Commission. (In peacetime, the yellow flag signifies "Plague on board.") This time, instead of freight, there were 300 sailors being sent to relieve the crew of the *Jeanne d'Arc,* the principal element of Admiral Robert's fleet. With American backing, this man held Martinique and Guadeloupe under Pétain's sway. For most of the passengers, Martinique was not a final destination. All of them had plans, and many had secrets. The physiologist André Mayer, of the Collège de France, traveling with his wife and son, hoped as I did to be able to teach again in America. Bloch, a porcelain manufacturer from Limoges, was counting on business contacts to set him up. A number wished to join forces with de Gaulle, either by going all the way to the United States or through channels in Martinique itself. This was the case with Corniglion-Molinier, who was to become de Gaulle's Aviation Minister. Some, no doubt, were simply leaving a continent unfit for human habitation. The monocle worn by an extremely taciturn passenger was enough to start a widespread rumor that he was

---

1   *Translator's note:* This is a pun for *Colis-Denrée,* i.e. package with foodstuff.

a follower of Hitler's. It was said that he smoked cigarettes stamped with a swastika.

There were too many passengers for a family like mine to have the use of a cabin all to ourselves. Cabins were assigned by sex, with informal arrangements enabling couples to enjoy privacy now and then. Alain, who was nine years old, spent nearly all his time with the sailors, with whom he became quite popular; they taught him how to tie knots.

As we passed through the straits of Gibraltar, the sailors confided that some of their number had jumped overboard in the hopes of being fished out by English ships so that they might join the Gaullist navy. At Casablanca, our ship put in to port for three days. We would have liked to visit Meknès or Marrakesh but were told that our stay in this port might be cut short, so we had to be content with a visit to Rabat and its old city. Alain has never forgotten the storks in Rabat, with their noisily chattering beaks.

The *Winnipeg* was not a fast ship. It took over two weeks to reach Fort-de-France. At that time, this was a beautiful, magnificently situated city, with low buildings surrounding a square where we drank one punch after another. We were enchanted by our visit to the market, with all its varieties of fish, bananas, and fruits of every color. Through a friend of my sister's, Dr. Bercher, I had obtained a letter of introduction to Des Etages, who was famous on the island for having killed a governor responsible for the death of his father. I do not know how much time he had had to spend in prison. His wife was French, and they received us with the utmost kindness, taking us swimming and sailing at the nearby beaches. Under cover of a small business run by his wife, he had organized the network by which volunteers were taken to the territory of Saint Lucia, whence they could go on to join de Gaulle's forces. Des Etages was eventually arrested and imprisoned under harsh conditions, I believe; after the liberation he served as Deputy from Martinique.

Like us, many of the passengers on the *Winnipeg* were seeking passage to the United States. Such transportation was difficult to find, costly, and uncomfortable. Luckily, after a few days, Admiral Robert observed that we were demoralizing his subjects by showing them that not all of France supported Pétain, and he ordered us to board the *Winnipeg* without delay. In this we fared better than the passengers of the next boat, the *Wisconsin,* whom the Admiral subjected to several weeks in a concentration camp. We next put into Guadeloupe for three days, which gave us a chance to tour the island, and we were then taken to the American territory of Saint-Thomas in the Virgin Islands, where at the end of February, after over a month at sea, we said our goodbyes to the *Winnipeg.*

Although Saint Thomas had been a Danish possession until fairly recently, we were prepared to find there a foretaste of the Coca-Cola culture. The heat was stifling, but there was a dearth of Coca-Cola; perhaps the American Navy had drunk it all – though by no means to the exclusion of other, more powerful, beverages. All I could offer Alain was a little tepid Pepsi-cola, in a bar peopled by tipsy sailors. To Alain's delight, one of them tried to enter the bar on horseback. A small American boat, the *Catarina,* took us to Puerto Rico just in time to obtain the last three bunks on a regular liner bound for New York. The sea was rather rough, and Eveline could hardly leave the cabin. When I took Alain to eat breakfast in the morning, we were shown to a table generously laid with pots of jam, where an American woman was already seated. Alain, accustomed to the penurious conditions on the *Winnipeg,* stared at the preserves with great round eyes, asking me, "Does all that belong to the lady?"

Chapter VII
The Americas; Epilogue

Once I read about some survivors of the Pacific war who spent many weeks adrift in a dinghy with only a pocket map to guide them. They finally drifted onto an island thousands of miles from where they thought they were. The key to their survival, they said, lay in always keeping a steady hand on the tiller despite the winds and currents, and in thinking of themselves as masters of their own destiny – an illusion, but this illusion was their salvation.

I, without a doubt more fortunate than these navigators, had reached the precise destination I had set my sights on throughout my peregrinations: New York. Once there, I thought my future was secure. The first order of business was to find out what type of position was in store for me at the New School.

A school official had been sent to meet me when the boat docked. His first step was to take me and my family to the Brevoort Hotel, a well-appointed place in the old style, which some older New Yorkers still remember with nostalgia. Located on majestic Fifth Avenue, it was not far from Washington Square, Greenwich Village, and also the New School.

Straightaway I found out that my position at the New School was fictitious: it had been invented for the sole purpose of qualifying me for a "non-quota" visa – a purpose it had not even accomplished. In fact, the salary I had been promised was to come from the Rockefeller Foundation, as part of a sweeping program to bail out French scientists.

There were certainly grounds for fearing that the Germans were to behave in France as they had in Poland the moment they invaded that unfortunate country, where they had already begun the systematic destruction of the intelligentsia. In France, the situation was somewhat different: even among the Jews, many intellectuals had no need of emigrating in order to escape the turmoil. Still, no one could know for certain, and the risks were only too real. This was the principal argument presented to the Rockefeller Foundation by Louis Rapkine, a brilliant young biochemist from Canada who happened to be in New York at the critical moment. He had drawn up and presented to the Foundation a list of French scientists, including both Jews and non-Jews, whom he thought it advisable to rescue from France as rapidly as possible. As soon as he found out – how, I do not know – that I was back in France in October of 1940, he added me to his list, and it was thus that I received the offer from the New School. I used to call him Saint Louis Rapkine. To this day I am touched when I remember the kindness that he showed me. After the liberation, he managed with some difficulty to find a position at the Institut Pasteur, only to die soon after of lung cancer. I was deeply saddened by his death.

Thus I found myself in New York, free of any obligations for the moment. As the salary of $2500 a year paid me by the Foundation was more or less sufficient for our living expenses, I was spared immediate financial worries, but I still needed to think about finding for the long term a more stable, and if possible better remunerated, position. Other mathematicians of my generation, to whom I did not think myself inferior, had succeeded in finding such positions. Ignorant as I was of how things worked in America, I was not particularly worried.

Another major concern was for my family in France. My parents and my sister had stayed on in Marseilles, and I did not think they would be safe there in the long run. A cousin of my father's, Blanche California Weill, was living in New York; her father, my father's first cousin, also an Alsatian but significantly older than my father, had emigrated in the nineteenth century and founded what was to become the largest department store in Bakersfield, California. His daughter Blanche was a bright and energetic child psychologist, one of the first women to earn a doctorate from the University of California at the beginning of the century. She generously offered to sign the sponsorship papers my parents and sister needed to obtain "quota" visas: these documents made her responsible for all their financial needs once they were in the United States. But even after this indispensable formality was cleared, there was a long waiting period. Furthermore, while my parents were eager to join me in America out of concern for their own safety and even more for that of my sister, she on the other hand by no means shared their eagerness. As a condition for coming she wanted my word that once here she could implement her plan – which she mistakenly thought I knew all about. This plan, to establish a corps of front-line nurses, was something she had discussed in great detail with the poet Joë Bousquet before leaving Marseilles.[1] De Gaulle's reaction when her plan was later proposed to him was, "She's crazy!" In fact she had conceived this plan above all as a way of sharing physically in the most dire suffering of the men engaged in combat. Obviously, I was unable to give her the assurance she wanted from me.

Fortunately, communication between Vichy France and the United States was still allowed, but naturally it was subject to censorship as well as to lengthy delays. Correspondence was not allowed with occupied France, so that Eveline had to rely on the good offices of friends in order

1   Cf. her *Correspondance avec Joë Bousquet* (Lausanne: L'Age d'Homme, 1982) and her *Ecrits de Londres* (Paris: Gallimard, 1957), 187-195.

to communicate with her mother; even so communication was slow and unreliable.

In all respects I believed my apprenticeship was now behind me. It did not take me long to find out how mistaken this notion was. I had arrived in New York on March 3, 1941. That very same month, the Rockefeller Foundation received a letter concerning me from Haverford College. Emphasizing my lack of experience in American teaching methods, and the detrimental effect this lack would have on my career in the United States, the letter offered to provide me (free of charge!) with the necessary experience by appointing me to a one-year teaching position in the department of mathematics at Haverford. As the Foundation was paying me, the position was unremunerated.

A "college" in the American sense[1] is an institution offering four years of instruction beyond the secondary education (which is often mediocre) provided in high schools. In the nineteenth century, three such colleges had been founded in the outskirts of Philadelphia by the Society of Friends ("Quakers"). The colleges were named after the townships in which they were located. The oldest of the three was at Haverford, and the other two were at Bryn Mawr and Swarthmore. This same religious sect had founded the city of Philadelphia and the state (officially "Commonwealth") of Pennsylvania in the seventeenth century. In 1941, Haverford was still an all-male institution and Bryn Mawr admitted only women, while Swarthmore was already coeducational. Each of these three colleges had an excellent reputation. At Haverford, the Quaker influence was still very much in evidence: a large number of professors and students regularly attended the Sunday religious observances, which consisted of silent meditation interrupted only by the words that inspiration might dictate to one or another of the participants. In fact it was only rarely, I believe, that one of these meetings ended without inspiration coming to one of those present; often, it was the professor of philosophy who spoke.

The Quakers are known not only for their moral virtues, but also for a sharp business sense – a trait that is probably not unrelated to the high regard in which they are held in American society. Haverford's offer did not appear overly generous to me, but the people at the New School strongly advised me to accept it. Without great difficulty I made up my mind to do so, and I did not come to regret this decision. I was told that it was a good way to get my feet wet in the American university system. On my first visit to Haverford, in early April, I ventured to suggest that a salary supplement would be welcome, and my boldness was rewarded with $250

1   *Translator's note:* the word *collège* in French designates certain institutions of secondary education.

above my yearly Rockefeller salary. For the rest, they treated me kindly, and the European in me could not help but feel at ease in the midst of the time-honored traditions of this college.

My arrival at Haverford was thus planned for the beginning of the school year the following September. In the meantime, my family and I settled in Princeton, where I knew most of the mathematicians. In truth, my recent past, and the gossip to which it had given rise, resulted in a rather cool reception from some of these colleagues. But I can hardly put into words the pleasure I felt in being reunited with my friends Siegel and Chevalley, of whom I had had only the most indirect news since 1939. Siegel had left Göttingen in the spring of 1940. Already resolved to make his way to America, he had arranged to be invited to Norway, where he had registered as a political refugee. From Bergen he had sailed on the last boat leaving for the United States – on the very same day that the Germans entered Oslo. It was chilling to imagine what would have happened to him if the German troops had got hold of him. As for Chevalley, about whom various rumors had circulated in France since the beginning of the war, he had duly reported to the French consulate and was told to stay on in Princeton, where Lefschetz had secured him an appointment as assistant professor. As soon as I saw Chevalley, he offered to share his office in Fine Hall with me. Day after day, arriving in this office, I would share with him the fruit of my reflections. As far as mathematics went, my greatest preoccupation was to work out in detail impeccable proofs of the results I had discovered while in the Bonne-Nouvelle prison a year earlier. Hermann Weyl, who had welcomed me with his usual kindness, offered to use his influence to have me put in prison again, since my previous stay in that type of establishment had had such a positive effect on my work.

Thus the spring and summer of 1941 passed. Chevalley, and especially his wife Jacqueline, were of great help in getting us settled in a house in downtown Princeton (the house is no longer there today), and in initiating us into the mysteries of shopping in America. It is true that supermarkets had not yet come into existence – or at least they had not yet made it to Princeton. During this time, Alain attended school, and before long he was speaking English better than his mother and I did. He was even luckier than I in finding gainful employment: in no time at all he was offered a job selling newspapers and magazines, a job for which his boyish French charm came in handy.

In autumn, we went off to Haverford, where our days would have been extremely happy had they not been ominously darkened by world events and by our worries about relatives in Europe. One of my colleagues, a man from an old Quaker family, and as such a convinced pacifist, returned from Europe where he had been as part of a delegation of Friends charged with assessing the situation. Their unanimous conclusion, predicting the United States' entry into the war in the near future, could be summarized in three words: "Arm like hell." One evening toward the end of November, a splendid *aurora borealis* lit up the sky over Philadelphia. This natural phenomenon is virtually unheard-of at such latitudes: was it some kind of omen? That evening police stations were inundated with telephone calls from anxious people asking whether the war was beginning here.

Apart from these worries, our life was peaceful enough. Alain was enrolled in school; Eveline was taking an American history course at Bryn Mawr. Altogether there were three of us in the Haverford mathematics department. My colleagues were friendly, and my teaching duties did not require any great effort on my part. The head of the department made no claim to being a serious mathematician, but he did have a rare sense of humor for an American. With his colleague, Carl Allendoerfer, a fruitful collaboration seemed to be in the offing. The mathematics professor at Swarthmore, a charming elderly Dutchman named Arnold Dresden, invited me to speak at his college about my recent work, for an honorarium of $500. This invitation not only provided me with a substantial supplement to my modest income, but was also a precious vote of confidence.

At the beginning of 1942, my colleagues told me what they thought was good news: a university in Pennsylvania was preparing to make me an offer. When the offer came, it was disappointing: it was only for an instructorship, and wretchedly underpaid at that – the type of position that would have been offered to a rank beginner in the field. Moreover, the fact that most of my salary would be paid by the Rockefeller Foundation had here again influenced the terms of the offer. Had it not been for my family, I would have been tempted to refuse the offer without regard for the consequences. At least it was explained to me that American university traditions include bargaining. Following this suggestion garnered me the title of "assistant professor" and a slightly higher salary. The feelers I had put out in the hopes of obtaining a more suitable appointment had come to nothing, and Eveline was pregnant. Somewhat later, after I had visited a university in the midwest, the friend who had had me invited there wrote: "It is out of the question that you would be hired here – not that your lecture didn't go over well, but for the three usual reasons: you're Jewish, you're a foreigner, and you're too good a mathematician for these people."

The 1942-43 academic year was about to begin, and with it a busy schedule for me, consisting of courses that I already knew would be depressing. In the form of a letter to Artin, I recorded my latest ideas on my research in algebraic geometry. At the same time, plans for my parents' voyage to the United States had taken shape: with my sister, they sailed from Marseilles on May 14. Early in July, after a long and difficult itinerary including Casablanca and Lisbon, they arrived in New York, where I met them. My sister immediately barraged me with her plans – the famous nursing corps, of course. She wished to waste no time in contacting Jacques Maritain, on whom she was counting to introduce her to Roosevelt; Roosevelt would approve the project and she would head for London to carry it out. When I told her things were not so simple, she was bitterly disappointed, but still she did not give up her plans.

In the meantime, she and my parents settled in New York, in a modest but well-situated apartment located on Riverside Drive, with a beautiful view of the Hudson. I moved to Bethlehem, a small city which is inoffensive enough except for its steel mills, which fortunately are located fairly far from the center of town. Here there were still many traces of the Pennsylvania Dutch, the German Mennonites who settled the area; their dialect was still spoken in the region. There was a peaceful cemetery in the downtown area, with only large rocks for tombstones, and a pretty rose garden adding to its charm.

Our daughter Sylvie was born on September 12, just before the end of summer vacation. To Eveline's great satisfaction, the hospital where she gave birth was located not in Bethlehem proper but in a neighboring village poetically named Fountain Hill: a birth certificate stating "Weil, born in Bethlehem" would have seemed a bit conspicuous to her, especially if the child had been a boy. While she was in the hospital, Chevalley came to lend me a hand with domestic duties, and Alain was highly amused to see us endlessly drying plates, discussing algebraic geometry all the while. Once Eveline came home, we felt the urgent need for help around the house. Unfortunately, with most men drafted, women were replacing them in the factories. With some difficulty we managed to retain a good person named Mary to leave her home in Nazareth and come help Eveline for a time with the most essential tasks. Some time later, Chern came to visit, having just arrived in Princeton from China. His visit, the beginning of a lasting friendship, was fruitful for both of us. I do not believe he was pressed into dish-washing duty.

My "teaching," if it can be called such, began at the end of September. The institution to which I belonged (a word that all too accurately describes my relationship with my employer) was graced with the noble title of "university"; but in fact, it was only a second-rate engineering school attached to Bethlehem Steel. The only thing expected of me and my colleagues – who were totally ignorant as far as mathematics went – was to serve up predigested formulae from stupid textbooks and to keeps the cogs of this diploma factory turning smoothly. Sometimes, forgetting where I was, I would get carried away and launch into a proof.

Afterwards, according to the well-established ritual, I would always ask, "Are there any questions?" Just as predictably would come the question: "Is that going to be on the exam?" My answer was ready: "You should know it, but it's not very important." Everybody was happy.

Thus the 1942-43 academic year passed. At least I was able to continue writing, though at a somewhat slower pace, my book on the foundations of algebraic geometry, the indispensable key to my later work. My parents, and then my sister, had of course taken the earliest opportunity to come visit our newborn. I was especially touched by my sister's emotion. She straightaway raised the question of the child's baptism, which she eagerly desired for reasons that I obviously could not share. Eveline had the same wish. Divorced, and living in sin in the eyes of the Catholic Church, she considered herself outside the church and, I am sure, felt no pangs about it; yet she still felt ties to her childhood religion, and on Sundays she often went to mass, where I would sometimes accompany her. All baptism meant to me was a few drops of water on a baby's head, as my sister pointed out, along with other *ad hominem* arguments. How could I possibly object? It did not take much to persuade me.

At the same time, the issue of Alain's religious allegiance was raised. During the period in question, in a city as provincial as Bethlehem, the label "No religious affiliation" entered in a school record was likely to be a strike against the child. Alain was the child of Eveline's first marriage, which had been celebrated in church; it seemed only natural to declare him Catholic. According to customary practice, it was time to entrust him to our local parish priests in order to prepare him for his first communion.

Most Catholic priests in the United States are of Irish or Italian descent, neither being known for their openness of mind. We had an illustration of this on Palm Sunday of the following year. Ever since we had attended Palm Sunday services at Saint Mark's in Venice, on our first trip to Italy together, this had been one of Eveline's and my favorite holy days. In 1943, Eveline went to church as usual, while I stayed home to look after Sylvie. Eveline came back outraged. The sermon had been more or less as follows: "How beautiful today's gospel is!," the priest had said; "How I wish I could speak of it at length! But we have more pressing problems. The collection from last Sunday was woefully inadequate. We have many expenses to cover..." It would be unfair to generalize from this one occasion; but words cannot describe Eveline's indignation that morning.

In any case, we were looking for something else. Fortunately, my sister had just met one of France's most distinguished Dominicans in New York: Father Couturier, who later achieved renown as the master mind behind the construction of Notre Dame de Toute Grâce in Assy. She explained our situation to him, and he agreed to take Alain's first communion in hand. To explain our request I told him, "I've heard a lot of bad things about the clergy in America." He replied ecclesiastically: "People exaggerate." Counter to my sister's expectations, he found it perfectly natural for Eveline to teach Alain his catechism. After a brief examination, he administered Alain's first communion mass in the French Dominican chapel in New York. Just before saying mass, he offered to let Eveline take communion as well. Surprised and touched, she nevertheless declined. As became evident later with the publication of his diary, he was imbued with the spirit of Vatican II. The mass was moving. I have never heard a priest pronounce the words "Ecce Agnus Dei" with such profound and visible faith.

During this time, my sister continued her attempts, if not to realize her plan (she had just about given up hope), at least to join the Gaullist organization in England. She was convinced that from there she would eventually succeed in entering France by parachute. She could not stand the thought of others suffering while she herself was living a sheltered existence. She did manage to meet André Philip, who agreed to enlist her in his Commissariat of the Interior.[1] In November she left for England. We were never to see her again.

The 1943-44 academic year began inauspiciously for me. Because of the war, many colleges and universities suddenly found themselves on the brink of bankruptcy. A large portion of their funding came from students' tuition, and the students had been devoured by the army. It quickly became evident that while troops were awaiting transport to Europe, and soon to Africa, they were condemned to otium in camps in the United States. To put this idle time to use and at the same time to occupy faculty members who were themselves idle, many of the troops were sent to colleges. The vast majority of these recruits knew nothing and did not care to increase their knowledge, but this fact made no difference. Under the aegis of the Army Science Training Program (known as the ASTP) they were sent to fill the void that the mobilization had created in the college dormitories and finances. I wound up having to spend fourteen hours a week spoonfeeding these poor boys the elements of algebra and of analytic geometry. They showed up in class in uniform, marshalled by a non-commissioned officer. Once, to obtain silence, I had him order them to attention. One day, one of them had a question: "I don't understand what x is." The question was far more profound than he suspected, but I did not attempt to explain why.

---

1   *Translator's note:* In the Free French organization, the word *"commissariat"* was used instead of the peace-time term "ministry."

I had at least hoped to have my summer free for a well-deserved vacation, which I expected to devote to the book that was still in progress, but this respite was denied me. The Rockefeller Foundation subsidy was about to expire, and there was no alternate solution in sight. At Rapkine's suggestion, I wrote a long letter to Warren Weaver, one of the Foundation's directors, asking whether he might not be able to use his influence at least to obtain an appointment for me at some other university where the scientific atmosphere was more breathable. He replied immediately with an encouraging letter, but this was soon followed by another in which he advised me to sit tight and wait out my discomfort. I found out later that in the interval between these two letters he had consulted Lefschetz, who had apparently taken my letter to Weaver as a reproach meant specifically for him. Lefschetz was reputed to be among those Jews who, to avoid any accusation of favoring their co-religionists, go to the other extreme and display patently anti-Semitic behavior. A conversation I had with him on this subject soon afterwards left me dumbfounded.

At least my life was free of worry as far as my immediate family was concerned. Eveline was in good health, and Sylvie was flourishing. The only problem in her case was finding the brand of condensed milk prescribed by the doctor, for during this period of occasional shortages this milk sometimes disappeared from the shelves for days at a time. My sister, who had arrived in England and was already in the service of André Philip, sent news regularly to my parents, and occasionally to me. My parents, still in New York, lived in the hope that they would be able to join her. When the Gaullist organization gained a foothold in Algeria, they counted on being reunited with her there, some day soon.

In fact, as we found out later, Simone – after desperate efforts to get herself sent on a mission to France, most certainly to her death – had had to be hospitalized in Middlesex Hospital in April. From here she had been transferred, already close to death, to the Ashford Sanitorium in Kent, where she died on August 24, 1943. She had done everything she could to keep her family from finding out about her condition, and she had succeeded.

Nothing, then, had prepared me for the telegram I received from a close friend, Madame Closon; to this day the telegram remains etched in my mind: "Simone died peacefully yesterday, she never wanted to let you know." Louis Closon, who had been in the United States when the war broke out, had joined the Gaullist organization early on. He distinguished himself in the resistance, earning recognition as a *Compagnon de la Libération*. I had met him in New York in 1941. In 1943, he was with his family in London, when he was not on a mission elsewhere; he and his wife were among Simone's small group of close friends. Madame Closon's

cable was so unexpected that I could not refrain from wondering if I should
believe it. I knew that Madame Closon had earlier suffered a nervous
breakdown. Of course, the cable was only too true.

How can I describe my grief? But I did not have the luxury of
indulging it; it was up to me to inform my parents, and I did not feel equal
to the task. Luckily I knew that Dr. Bercher, a friend of my sister's, was in
Philadelphia at the time. He had played a significant role in exposing the
darkest aspects of the French colonization of Indochina; both he and my
sister had contributed to an independent leftist periodical, *La révolution
prolétarienne*. A doctor in the Merchant Marine, he had spent a large part
of his life at sea: it was in this capacity that he came to know Indochina,
and that he now happened to be in Philadelphia. I succeeded in reaching
him by telephone. Without hesitation, he accepted the gloomy mission of
accompanying me to break the news to my parents.

When we met in New York, we decided we should soften the blow
we were about to inflict upon my parents. We had a mutual friend, himself
greatly distressed at the news, telephone them as if he had just heard
worrisome rumors about the state of my sister's health. Meanwhile, we
headed for my parents' apartment. They wanted to send a cable asking for
news of my sister. My father went down to the lobby, where the only public
telephone in the building was located. Bercher and I followed him there,
telling him rather clumsily, "It's no use." He understood instantly. Perhaps
for the first time in his life, he began to cry. "Our poor little Simonette," he
said (this had been his name for her when she was little); "she loved us so
much." Then, thinking of my mother, he said, "How can we break it to
her?" After hearing the news, she talked of committing suicide, the two of
them together. It took quite some time to get her to let go of this notion.

The rest of 1943 was rather bleak for me, and even bleaker for my
parents. Thanks to Eveline, my home life was perfectly happy, but this was
my only source of joy. There were no prospects on the horizon for
improving my work situation. In January I sent an S.O.S. to Hermann
Weyl, writing: "Prostitution consists in diverting something of high value
to base uses for mercenary reasons; this is what I have been doing these
two years." I announced my decision to resign my current position,
whatever the consequences, and I asked his help in obtaining a means of
livelihood for myself and my family. He replied that he understood how I
felt, that many others besides me were in a similar situation, but that "some
are more sensitive than others to organized stupidity," and that he would
see what he could do for me, hoping for the best. He did in fact contact Dr.
Moe, secretary of the Guggenheim Foundation, who ran the Foundation
virtually single-handedly. Dr. Moe was understanding and sympathetic.

According to the Foundation's regulations, it was already too late to apply for a grant for the current year – but in no time I was awarded a grant.

In the meantime another possibility had cropped up. In New York, I had met the sociologist Lévi-Strauss, and we had hit it off quite well. I had solved for him a problem of combinatorics concerning marriage rules in a tribe of Australian aborigines. He had taught for several years at the humanities section of the University of São Paulo in Brazil. In 1944, he introduced me to the geneticist Dreyfus, the dean of that institution, who was then on a research trip to the United States. The University of São Paulo, only recently founded, had at first chosen French scholars to teach disciplines in the humanities and Italians for the sciences. Because war had been declared between Brazil and Italy, the Italian professors had been obliged to repatriate; there was thus a chair in mathematics that needed to be filled. To my great good fortune, Lévi-Strauss and Dreyfus thought of me for the position. My official appointment to it followed shortly.

But it was not enough to be selected on the Brazilian end. Because of the state of war, no foreigner was allowed to leave the United States without an exit visa issued by the Immigration Service. I therefore applied for a visa for myself and my family, including my parents, who did not want to be left behind in New York. To my great surprise, my request was denied. Dreyfus intervened on my behalf with the State Department, but to no avail. Neither I nor Dreyfus was ever told what was behind this refusal. Most likely, in my view, was that the American cultural attaché in Brazil intended to reserve this position for a compatriot, and had made his wishes known in Washington.

Other worries began to hound me. Eveline, who had had tuberculosis before the war, but who we thought was finally cured, showed signs of a relapse. On the advice of her doctors, I took her to a sanatorium. Alain was left with friends in Princeton, and Sylvie with my parents. We left Bethlehem with no regrets.

Meanwhile, my spate of ill luck was coming to an end. For assistance in obtaining a visa, I sought out Aydelotte, the director of the Institute for Advanced Study at Princeton. I told him that it was improper for me to be kept against my will in a country that had no place for me. I wonder whether he invoked this argument with the Immigration Service, who then and for a long time after were known for their narrowness of mind. No matter: I was granted my visa. At about this time I found out that my friends the Zariskis would be in São Paulo as well. It seems to me that I had suggested this name to Dean Dreyfus in case I should be unable to go to Brazil. Was this arrangement a compromise to satisfy the American cultural attaché? I had no way of knowing. I could only rejoice at the prospect of being reunited with these friends in Brazil.

In the meantime, Eveline had been released from her sanatorium. We made ready to leave at the end of the year. My grant from the Guggenheim Foundation was sufficient to meet our needs until our departure. I asked Dr. Moe whether I should submit my resignation effective on the date of our planned departure from the U.S. Being familiar with South America, he told me, "I will suspend your checks as of that date, but don't resign until you get to Brazil. Once you are there, if everything works out as planned, then you can send me your resignation."

While awaiting the time of departure, Eveline and I moved to Swarthmore to be with Arnold Dresden, the old Dutchman whose friendship had already proved precious to me. One of his colleagues at Swarthmore was the renowned British poet W. H. Auden, who readily agreed to read over with me the introduction I was writing for my book on algebraic geometry. Thus the day before we sailed from New Orleans, I was able to send the complete manuscript to the American Mathematical Society, in the expectation that they would publish it. This they did, after stipulating that I excise a somewhat bitter sentence from the preface.[1] In truth, I was expecting this, and had no problem complying with it. Apart from this detail, the book was accepted. I corrected the page proofs in Brazil in the following months.

This time, everything turned out as planned. We (Eveline, Alain, Sylvie, my parents, and I) were to sail on the *Rio Tunuyan,* a little Argentine liner, or more exactly a French liner that happened to be laid up in Buenos Aires during a brief period when France was theoretically at war with Argentina. Argentina had taken advantage of the situation to seize the ship, change her name, and supply her with an Argentine crew. A year later, the Argentines were forced to return the boat to France. The voyage to Rio was uneventful but extremely long, for the war, still in progress, made it necessary for us to follow an unwonted route along the coastline. Sylvie quickly became a favorite among crew and passengers alike; so did Alain, for that matter, especially after performing a puppet-show which succeeded in coaxing a smile from the dour-faced captain.

A young officer on board, who enjoyed playing with Sylvie, confessed to us his disappointment at the brevity of the ship's lay-up in New Orleans. It was the first time ever that he had left Argentina, and he had been looking forward to it eagerly. He had met a girl there and persuaded her to come back to his hotel room, where, he said, she had

---

1   This sentence was the following: "[I owe an enduring debt of gratitude to...] and to those few who strove to liberate me from the distasteful and humiliating duties of a job which my position as a refugee in the United States had compelled me to accept."

permitted him "all but the main thing." This she refused, saying "I can't, I'm Catholic." "But, good God, so am I!" answered the poor fellow; thus ended the poor man's adventure.

All this time, since the French government had returned to Paris, my friends, particularly Henri Cartan, had not been oblivious to my fate. I had been dismissed from my university position in 1939. In 1944 my friends had me "reintegrated," to use the administrative jargon, in the *cadres* of the French university system. As I was later told by an employee of the Ministry of Education, for all its faults the French administrative machine has at least one advantage: it can go backwards in time, like a time machine. Henceforth, in the eyes of the system, I had never been dismissed. I was simply temporarily detached (*"détaché"*) to Brazil. As a compatriot and colleague in Brazil once said to me, as we left the cultural attaché's house: "You see, Weil, there are two kinds of people in the world: the *attachés* and the *détachés."*

In Rio, then, the machine had notified the ambassador of my imminent arrival. The cultural attaché welcomed us courteously and dispatched us to São Paulo, where Dean Dreyfus was expecting us.

I arrived in São Paulo in January, 1945, in the middle of summer and of summer vacation. Dreyfus explained that my salary would accrue as of the time I arrived, but would not actually be paid out until after the medical examination required by law for all employees, even temporary ones, of the Brazilian government. This included me, as the University of São Paulo was a state university. As this formality could take months, Dreyfus would advance me money of his own to tide us over. This salary, along with the supplement the embassy paid me in my capacity as *détaché,* left me quite comfortably off, especially since an article of the Brazilian constitution exempted priests, professors, and, I believe, journalists from income tax. This article has since been repealed.

Naturally one of our first concerns was finding a place to live. My French colleagues, in particular the geographer Pierre Monbeig (the son of my old eighth-form professor at the Lycée Montaigne), were of great help to us in this matter. What a contrast with our gloomy apartment in Bethlehem! A modest but comfortable house, a pleasant garden, a big mimosa, and roses blooming all year round, often visited by hummingbirds (in Brazilian Portuguese, *beija-flor* or "kiss-blossom"): what more could we want? In addition the rent was surprisingly low. My colleagues told me why: the owner of the house was a German who had headed the Nazi cell in São Paulo. In Brazil rents were strictly regulated by law. Though few landlords abided by these rent controls, ours thought it wise to observe them, so as not to call attention to himself. We had an excellent relationship with him.

It was in March, I believe, that courses began. I had been appointed as a *contratado,* or temporary professor, in advanced analysis. At that time, the mathematics department and library were located in a charming villa in a pleasant neighborhood. I would go to work either on foot or by tram (called the *bonde* in Brazilian Portuguese). A servant-girl was employed to clean the department and especially to serve every professor, both before and after classes, as many *cafezinhos* (small cups of Brazilian-style coffee) as he wished. When I visited Brazil in 1966, I observed that this excellent custom had been discontinued, and that the mathematics department was now housed with the university's other departments in huge buildings on a sprawling campus devoid of character.

My teaching duties were quite light. My predecessor, Albanese, a geometer of some distinction, had assembled a departmental library that was exceptionally good, especially in his own field, algebraic geometry. It was excellent for my purposes. The city of São Paulo has a large population of Italian descent, and many people speak fluent Italian; besides, Italian and Brazilian (that is, Portuguese spoken with a softer accent than in Portugal) are fairly close. It is possible that Albanese conducted his courses in Italian; I believe this is what Zariski did, at least during the early part of his stay. Before coming to America, Zariski had lived for a long time in Rome, where he had married an Italian. As for me, I started out teaching in French, and the second year I switched to Brazilian, which I picked up easily – though not as easily as Eveline, who learned from conversing with the maid we hired soon after arriving (a luxury we had long since done without).

As soon as Paris was liberated in 1944, mail service started up again. Eveline and her mother were thus able to correspond without hindrance, and I resumed my correspondence with Bourbaki. Before the liberation, I had received a report on only one congress: this was the miniconference[1] held in September, 1943 at Liffré, not far from Rennes, in a country inn where tasty and above all plentiful food was still to be had. This congress, attended by Cartan, Delsarte, and Dieudonné, had been especially productive. The problem arose of sending the conclusions to Chevalley and to me. One copy, sent to our colleague and friend Georges de Rham and forwarded to me from Switzerland, never turned up. Another, sent to London by clandestine Gaullist mail pouch, somehow reached New York, by which time it had lost its address. By mere chance it was seen by the physicist Francis Perrin, who thought he recognized Bourbaki's style and sent it straightaway to me. Once Paris was liberated and I was in Brazil, such detours became unnecessary, and our discussions were resumed as usual.

---

1   In French: *"congrès croupion"*, or "rump congress".

Soon my Parisian friends did even better. They obtained on my behalf the rare favor of a travel warrant flying me to Paris, all expenses paid, so that I could get back in touch with Bourbaki and with France. The French consulate notified me of this opportunity in June. Flights connecting France and South America had by then been reinstated. Thanks to the consulate, I showed up for the flight from Rio equipped with a small bag and a kilo of café that would, I was sure, be most welcome in Paris. As it happened I did not fly out that day: I had not been advised of the necessity of a yellow fever vaccination – I suppose because of the layover in Dakar. I had not been vaccinated, and orders from Paris were strict. The embassy managed to get me out on an American military flight. Thus it was that, "with a cargo of Weil and coffee," (to quote more or less the Bourbacchic write-up), the plane that transported me set me down, at Le Bourget I believe, on June 20, 1945. An impromptu Bourbaki congress was held immediately, recorded in the Bourbaki annals as the "coffee congress."

Life in Paris in that summer of 1945 was far from simple. The very first day, I had to obtain the indispensable ration cards: the bread card, the meat card, the "oils and fats" card, all of which remained in use until 1948. The apartment on the Rue Auguste Comte had been occupied by the Germans, who had stripped it of everything that could be carried away: they had even tried, unsuccessfully, to carry away the bathtub. There were bullet marks on the inside walls from the fighting that had taken place in the neighborhood. One of my first tasks was to borrow a mattress and blanket from some friends and to take them to my place on a handcart borrowed from neighbors. Contrary to all expectations, I found a Jewish couple who had been friends of ours living safe and sound in the apartment they had lived in before the war. They had spent the war years fairly peacefully under assumed names in a village in Savoy. Both doctors, they had access to sources of food not available to everyone; they kindly invited me to share their breakfast for as long as I stayed. For the rest, restaurants were few and far between, but the Ecole Normale supplied a good deal of my nourishment, and of Bourbaki's as well during the congress.

Already before the war, Bourbaki had begun to take on new participants on a trial basis – we called them "guinea pigs." In most cases, they became full-fledged members. This time the *Normalien* Samuel was promoted to the rank of guinea-pig, and was also appointed keeper of the coffee. After preparing it each day in his study-room, he would then secure this rare and precious substance under lock and key. Our meetings took place at the Ecole Normale, and Samuel was the official recorder. His report was embellished with a sonnet in imitation of Mallarmé's *Cygne,* ending with the following tercet:

Bourbaki Congress at Pelvoux-le-Poët (1951). Right to left: A.W., H. Cartan, J.-P. Serre, J.-L. Koszul, J. Dieudonné, etc.

Il contemple étonné, comme enivré d'un philtre
l'adhérence, un manteau qu'il n'a jamais compris
que vêt, sur un compact, immobile, le FILTRE.[1]

The first visit I had paid upon arriving in Paris was to Henri Cartan. By a happy coincidence, I found myself at his home just at the moment his father was returning from Moscow. This was not even the day he was expected home. A delegation of French Academy members had gone to the Soviet Union to reestablish contact with their Soviet counterparts. Elie Cartan had taken with him the first Bourbaki publication (the Russian mathematicians had not yet heard of Bourbaki) as well as my book on integration in groups. I had not seen him since the day of my appearance before the military tribunal in Rouen in 1940, when he had testified on my behalf, and when I had been able neither to speak nor even to shake hands with him. In 1945 he seemed hale, though his health was soon to fail. "Ah, Monsieur," I said to him, "I hope I may I kiss you" – and I threw my arms around him. He

---

1 He gazes, amazed, as if intoxicated by a philter, at the adherence, a cloak he has never understood, vesting, on a compact space, motionless, the filter.

must have been rather taken aback but, with his usual kindness, he submitted graciously.

Except for a few traces of the battle for liberation, Paris was still miraculously intact for the most part. Also miraculously, the streets were nearly free from traffic. Will there ever be another opportunity to walk about as freely as it was possible to do then? The only cars to be seen on the streets were a few American military vehicles and some private automobiles equipped with gas generators. The air was breathable as never since. In July of 1945, one room of the Louvre had just been re-opened to the public. This was the *Salon Carré,* where the curators had assembled a selection of the most famous paintings, which had been safely stored in the château at Chambord during the war. It was said that a fire due to a worker's negligence had nearly sent the whole château up in flames and with it the Louvre collection hidden in the basement; fortunately, the fire was brought under control in time. Entering the exhibition room I saw an elderly gentleman, rooted to the spot in front of Rembrandt's *Bathsheba,* repeatedly exclaiming, "Bathsheba! My precious Bathsheba! I never thought I would see you again!" I was tempted to say the same a few days later when I passed the Cathedral at Chartres on the way to Parcé, where Eveline had sent me with news for her mother. The train to Parcé, which is the Le Mans train, passes Chartres with a good view of the cathedral. Eveline and I have never seen it without being deeply moved.

My return trip to Brazil was more difficult, and above all longer, than the trip from São Paulo to Paris had been. This was the time the Americans were transferring their entire military operation from Europe to the Pacific. All flights heading west were reserved for military purposes, and the French Ministry of Education, responsible for sending me back to Brazil, could not obtain priority status for me. I had to take the train to Lisbon and from there take a Portuguese ship for the excruciatingly slow voyage back to Rio. On August 7, we put in at Curaçao which is, or at least was, a Dutch colony. I went ashore and bought a newspaper. Suddenly I found myself understanding Dutch: the paper announced that the first atomic bomb had been dropped on Hiroshima.

* * *

Let this explosion signal the end of these memoirs. I had by this time arrived, or nearly so, *"al mezzo del cammin di mia vita."* Since that return to Brazil, I have led the peaceful life of a mathematician, now and then illuminated by the joys of scientific discovery, but also by the pleasure of my many travels (most often in the company of Eveline) and by the contemplation of masterpieces the world over. Should I go on to speak of

Greece, Japan, or China – or, closer to home, of Poitou or Burgundy? Let the reader's imagination linger over these tantalizing names, to which I might add (as does Homer in his catalogue of Nereids), "... and many more besides."

But doesn't such a tale call for an epilogue? My stay in Brazil, with all its many pleasures, could not last forever. The chair I was occupying had sooner or later to be reclaimed by a Brazilian mathematician. Besides, despite Zariski's presence in São Paulo in 1945, and Dieudonné's in 1946 and 1947, it was impossible not to wish for a more stimulating scientific milieu. My Parisian friends thought it possible to arrange for my appointment to the Collège de France when Lebesgue's retirement left a chair vacant, but this plan did not materialize. Fortunately for me, my friend Marshall Stone had just been named head of the department of mathematics at the University of Chicago, with the mission of revamping the department from top to bottom. He offered me a chair, which I accepted. In the fall of 1947 I went to Chicago with Eveline and our two daughters, the younger of whom, Nicolette, had been born on Saint Nicholas' day, December 6, 1946. I did not leave the University of Chicago until 1958, when I went to the Institute for Advanced Study in Princeton. In 1976 I retired from the Institute. And now, though not without a tinge of melancholy, I can at last take leave of all these memories.

Index of Names